the STUFF of LIFE

the STUFF of LIFE

A graphic guide to genetics and DNA

Written by MARK SCHULTZ

Art by ZANDER CANNON and KEVIN CANNON

A Novel Graphic from Hill and Wang

A division of Farrar, Straus and Giroux

New York

Hill and Wang

A division of Farrar, Straus and Giroux

18 West 18th Street, New York 10011

This is a Z FILE, INC. Book

Distributed in Canada by Douglas & McIntyre Ltd.

Printed in the United States of America

Published simultaneously in hardcover and paperback

First edition, 2009

Library of Congress Cataloging-in-Publication Data

Schultz, Mark, 1955–

   The stuff of life : a graphic guide to genetics and DNA / written by Mark Schultz ; art by Zander Cannon and Kevin Cannon.— 1st ed.

       p. cm

   Includes bibliographical references and index.

   ISBN-13: 978-0-8090-8946-8 (hardcover : alk. paper)

   ISBN-10: 0-8090-8946-7 (hardcover : alk. paper)

   ISBN-13: 978-0-8090-8947-5 (pbk. : alk. paper)

   ISBN-10: 0-8090-8947-5 (pbk. : alk. paper)

   1. Genetics—Comic books, strips, etc. 2. DNA—Comic books, strips, etc. I. Cannon, Zander. II. Cannon, Kevin. III. Title.

QH437.S348 2009

576.5—dc22

                                                                                        2008033829

Editor: Howard Zimmerman

Design: Zander Cannon and Kevin Cannon

Science Consultant: Dave C. Bates

Production Assistant: Elizabeth Maples

www.fsgbooks.com

1 3 5 7 9 10 8 6 4 2

For Grace

--Mark Schultz

This book, and every book forever and ever, is for Julie

--Zander Cannon

For R.A.

--Kevin Cannon

The Stuff of Life represented a significant step into new territories for me. As such, it required an unusually high degree of consideration and cooperation among all those involved. My thanks to Zander Cannon and Kevin Cannon, who took the script and made it much more than the sum of description and text; to Dave Bates, our scientific consultant, who astutely kept us on the razor's edge between factual correctness and format necessity; to Howard Zimmerman, who conceived this project and wrangled it through, wielding a deft and thoughtful editorial hand; and to Thomas LeBien, whose vision and open-mindedness made it all possible.

--Mark Schultz

We would like to thank the many generous individuals who have helped make this book a reality. Everyone involved in this project has been extremely enthusiastic and exceedingly helpful to us from day one. Thanks to Mark Schultz for handing us a script that was truly a pleasure to draw, to David Bates for his knowledge and commitment to making sure we got our science facts straight, and to Howard Zimmerman for being our fantastic editor, sounding board, liaison, and cheerleader. We all worked hundreds--sometimes thousands--of miles apart, but it never felt that way. Finally, a huge thanks to Thomas LeBien at Hill and Wang, the man responsible for realizing this gorgeous book and for taking a chance on this unique way of teaching science.

--Zander Cannon and Kevin Cannon

CONTENTS

the STUFF of LIFE

the STUFF of LIFE

A graphic guide to genetics and DNA

WORDS Mark Schultz

Zander Cannon
Kevin Cannon ART

A DETAILED TRANSLATION OF THE REPORT OF **BLOORT 183**, INTERPLANETARY BIOLOGIST OF THE GLARGAL ROYAL SCIENCE ACADEMY, TO HIS SUPREME HIGHNESS **FLOORSH 727**, CONCERNING THE UNIVERSAL REGENERATIVE AND REPRODUCTIVE STRATEGY OF LIFE ON THE PLANET **EARTH**, SATELLITE OF SOL.

AN EDIFYING AND SOMETIMES SHOCKING EXPLORATION OF THE GREAT **DIVERSITY, TENACITY,** AND **INVENTIVENESS** OF LIFE ON THAT WORLD, AND A CRITICAL LOOK AT THE DOMINANT SPECIES' EVOLVING MASTERY OF GENETIC MANIPULATION.

FEATURING ASSORTED VISUAL AIDS AND SIDEBARS.

THIS IS **EARTH**, AS IT APPEARS TODAY.

IT'S A MEDIUM-SIZE PLANET WITH ENOUGH MASS TO HOLD A VIABLE ATMOSPHERE, BUT NOT SO MUCH AS TO CRUSH ASPIRATIONS...

EARTH'S EARLY HISTORY WAS FAIRLY TYPICAL OF SUCH CELESTIAL BODIES: FIVE BILLION YEARS AGO IT WAS NO MORE THAN A RIVER OF DUST AND ROCKY FRAGMENTS CIRCLING **SOL**, ITS SUN.

OVER ENOUGH TIME, EVEN **DUST** CLOTTING TOGETHER CAN GAIN ENOUGH MASS TO CREATE A GRAVITY CHAIN REACTION.

4.6 BILLION YEARS AGO

THROUGH THE ACTIONS OF GRAVITY AND ACCRETION, EARTH WAS BORN--

--A SEETHING, MOLTEN WORLD COALESCING UNDER THE IMPACTS OF LARGER AND LARGER ADDITIONS.

THE RADIOACTIVE DECAY OF ISOTOPES DEEP WITHIN THIS TORTURED BALL CREATED AND STOKED A CONSTANT FURNACE AT ITS CORE.

4.3 BILLION YEARS AGO

THE SURFACE WAS CONTINUALLY PELTED WITH A RAIN OF METEORITES BEARING EXOTIC ELEMENTS-- MOST NOTABLY **CARBON**, AS WE'LL COME TO SEE.

THEN A LONG PERIOD OF COOLING BEGAN, WITH THE YOUNG PLANET'S BOWELS DISGORGING THE **GASES** AND **STEAM** THAT CONTRIBUTED TO ITS OCEANS AND ITS NASCENT **ATMOSPHERE**--

4.1 BILLION YEARS AGO

--SNUGLY HELD IN PLACE BY EARTH'S GRAVITY-GENERATING MASS.

ALL THESE FACTORS, PLUS EARTH'S CONGENIAL DISTANCE FROM SOL, ITS ROTATION, AND THE SUBTLE TILT OF AN AXIS THAT DISPERSES SOLAR HEAT AROUND ITS SURFACE...

...CREATED THE PERFECT CONDITIONS FOR THE EXTRAORDINARY PHENOMENON KNOWN AS...

LIFE!

3.9 BILLION YEARS AGO

THE TRANSMUTATION OF BASE ELEMENTS INTO COMPOUNDS JUMP-STARTED BY A SUPPLY OF ENERGY INTO SELF-PERPETUATING ORGANISMS! THE CAPACITY TO GROW, TO REACT TO STIMULI, TO REPLICATE, RENEW, AND REPRODUCE!

EARTH'S FIRST LIFE WAS PROBABLY AKIN TO THE SULFUR-EATING **HYPERTHERMOPHILE** BACTERIA THAT STILL EKE OUT THEIR EXISTENCE IN THE MOST EXTREME ENVIRONMENTS, DOING QUITE WELL IN THEIR GEOLOGICALLY ACTIVE NICHES WITHOUT THE BENEFIT OF SUNLIGHT OR OXYGEN.

BUT THE ORGANISMS THAT POINTED TO THE FUTURE OF LIFE ON EARTH WERE THE **CYANOBACTERIA** THAT INVENTED **PHOTOSYNTHESIS**-- THE ABILITY TO CREATE THEIR OWN ENERGY FROM SUNLIGHT.

FOR MILLIONS OF YEARS VAST COLONIES OF THEM-- MATTED AND STACKED INTO **STROMATOLITES**-- DOMINATED THE PLANET...

3 BILLION YEARS AGO

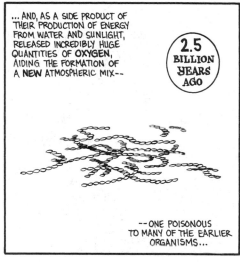

... AND, AS A SIDE PRODUCT OF THEIR PRODUCTION OF ENERGY FROM WATER AND SUNLIGHT, RELEASED INCREDIBLY HUGE QUANTITIES OF **OXYGEN**, AIDING THE FORMATION OF A **NEW** ATMOSPHERIC MIX--

2.5 BILLION YEARS AGO

--ONE POISONOUS TO MANY OF THE EARLIER ORGANISMS...

... BUT OFFERING FANTASTIC OPPORTUNITIES FOR BOLD, NEW EXPERIMENTATION. SOME NOVEL PROTIST ENVELOPED AND CAPTURED BACTERIA, AND THEN FOUND THAT THEY ALL COULD WORK TOGETHER-- **SYMBIOSIS!**

2.1 BILLION YEARS AGO

THE BACTERIA EVOLVED INTO COOPERATIVE ORGANELLES, AND **COMPLEX CELLS** WERE BORN!

COMPLEX CELLS BEGAN JOINING TO FORM **MULTI-CELLULAR ORGANISMS.** AT SOME POINT NOW LOST IN TIME, SOME OF THEM EVOLVED A REPRODUCTIVE METHOD CALLED **SEX,** WHICH ALLOWS FOR THE CHARACTERISTICS OF TWO PARENT ORGANISMS TO BE PASSED IN RICH, VARIABLE COMBINATIONS TO A NEW GENERATION, AND THINGS STARTED EVOLVING MUCH FASTER!

1.2 BILLION YEARS AGO

MULTI-CELLULAR ORGANISMS CONTINUED TO EVOLVE AND DIFFERENTIATE INTO THREE BROAD CLASSIFICATIONS:

PLANT   ANIMAL   FUNGUS

PLANTS ARE AUTOTROPHS -- THEY PRODUCE THEIR OWN FOOD BY PHOTOSYNTHESIS, THE SAME TECHNIQUE PIONEERED BY THE ANCIENT CYANOBACTERIA.

LARGELY IMMOBILE, THEY ARE IN CONSTANT COMPETITION FOR THE BEST LIGHT CONDITIONS.

FUNGI AND ANIMALS ARE HETEROTROPHS -- CONSUMERS OF FOOD SOURCES PRODUCED BY OTHERS.

ANIMALS THEMSELVES ARE FURTHER DIVIDED INTO TWO GROUPS --INVERTEBRATES, WHICH, LIKE THIS LOVELY, DELICATELY FORMED SEA CUCUMBER, ARE NOT ENCUMBERED BY A SPINAL CORD...

...AND VERTEBRATES, WHICH HAVE DEVELOPED A SPINAL CORD. IN SOME CASES, VERTEBRATES HAVE EVEN DEVELOPED ENTIRE INTERNAL SKELETAL SYSTEMS, WHICH COME IN HANDY, I GUESS, IF YOU WANT TO LIVE ON LAND AND ARE OBSESSED WITH SIZE.

ACTUALLY, SOME OF THE MOST INNOVATIVE AND EXCITING RECENT EVOLUTIONARY ADAPTATIONS HAVE BEEN EXPRESSED BY THE WARM-BLOODED VERTEBRATES KNOWN AS BIRDS AND MAMMALS...

...WITH THE MAMMALIAN SUBCLASS CALLED PRIMATES DEVELOPING A SPECIALIZATION IN THAT PARTICULARLY FASCINATING CHARACTERISTIC --

INTELLIGENCE!

SO, HERE WE HAVE THE PRIMATE, **MAN**, IN HIS CURRENT STATE.

MAN IS THE LONE ORGANISM ON EARTH THAT HAS DEVELOPED A **SCIENTIFIC** AND **TECHNOLOGICAL** CULTURE AND IS THUS OF PARTICULAR INTER--

Excuse me ...?!

Let me get this straight-- the creature you called the "sea cucumber"?

The sea cucumber **ISN'T** Earth's dominant organism?

WELL, I WAS CONFUSED AT FIRST, TOO, YOUR ROYAL INQUISITIVENESS. I MEAN-- THE CUKES **DO** LOOK SOMETHING LIKE US.

BUT THE AMAZING TRUTH IS-- THEY ARE AS DUMB AS HAMMERS.

IN FACT, NO EARTHLY INVERTEBRATE HAS EVER DEVELOPED INTO A GLOBALLY DOMINANT SPECIES, AS WE DID HERE ON GLARGAL.

NO, SIRE, ON EARTH IT WAS THE VERTEBRATES ...

...LAND CREATURES PAINFULLY BATTLING GRAVITY WITH WHAT MUST BE TERRIBLY ITCHY INTERNAL SKELETONS...

...THAT EVOLVED INTO THE PLANET'S TOP PREDATOR.

So you're telling me that VERTEBRATES--

--hairy, bipedal VERTEBRATES...

YES. AND THEY ARE NOW DOING REMARKABLE RESEARCH INTO THEIR BIOLOGICAL COMPOSITIONS AND ORIGINS, AS WELL AS INTO--

OKAY--let's... let's hold up a moment, Bloort,...

... I need some time to **DIGEST** this.

I mean-- I never considered...

...I never GUESSED that our counterparts on Earth might be so completely-- ALIEN.

Hairy, bipedal, land-dwelling vertebrates-- HMMPH.

Kush 1290-- are we sure that this... HUMAN SPECIES offers anything WE can use?

AN INSIGHTFUL QUESTION, MOST CAUTIOUS ONE. BUT, AS OUR GENETIC CRISIS HAS BECOME INCREASINGLY DIRE, I BELIEVE WE MUST LOOK TO OTHER SYSTEMS FOR HELP.

YES-- THE INFORMATION WILL BE A BIT CHALLENGING. AS YOUR SCIENCE ADVISER, I URGE YOU TO EXTEND A LITTLE TRUST...

...AND I'M SURE YOU WILL LEARN A GREAT DEAL FROM BLOORT 183. HE'S ONE OF OUR FINEST SCIENTISTS-- AN EXCELLENT OBSERVER AND THOROUGH RESEARCHER.

THERE WAS GOOD REASON WE TRANSMITTED HIM TO EARTH...

I MAKE NO ASSUMPTIONS, O MAGNIFICENT CORPULENCE. ALL MY RESEARCH IS BACKED BY EVIDENCE, PRESENTED HERE THROUGH THE SIMULACRON. AND I DO BELIEVE WE CAN LEARN MUCH FROM THE EARTHLY ORGANISMS...

Of course-- very well. You'll pardon me if my concern over the Royal Family's growing... DIFFICULTIES ...with the HERITABLE DISORDER have made me snappish. Of course I trust Bloort. We expect that what he has learned can only HELP us master our genetic TROUBLES.

It's just, frankly, I'm a bit disappointed. The sea cucumber was so much more RELATABLE.

I THINK, SIRE, THAT IT'S IMPORTANT FOR US TO REALIZE THAT THE SIMILARITY WE SEEM TO SEE BETWEEN THE EARTHLY SEA CUCUMBERS AND OURSELVES IS **DECEIVING.** IT IS A COINCIDENCE OF SEPARATE EVOLUTIONS MOLDED BY SIMILAR ENVIRONMENTAL FACTORS.

FOR INSTANCE, WE BOTH LIVE IN FLUID ENVIRONMENTS. MORE IMPORTANT, THOUGH: WE COME FROM **TWO DIFFERENT PLANETS.**

SO, **DESPITE** VISUAL SIMILARITIES, TWO DIFFERENT PLANETS + TWO DIFFERENT ORIGINS = NO BIOLOGICAL SIMILARITIES WHATSOEVER. ABSOLUTELY **NOTHING** IN COMMON.

TRUST ME, SIRE. YOU WOULD NOT RELATE WELL TO AN EARTHLY SEA CUCUMBER.

BUT-- TO GET US BACK ON TRACK -- DESPITE THE **APPARENT** DIFFERENCES BETWEEN SEA CUCUMBERS AND THOSE HAIRY MAMMALIAN BIPEDS--

-- THE FACT IS THAT **THEY** ARE REALLY NOT ALL THAT DIFFERENT!

AT ITS MOST FUNDAMENTAL LEVEL, ALL EXPRESSIONS OF LIFE ON EARTH ARE LINKED **VERY** CLOSELY TOGETHER. THEY SHARE **ONE COMMON ANCESTOR.**

AND THE ASTONISHING EVIDENCE LIES IN **EVERY** CELL THAT **EVER** LIVED ON THAT PLANET: EVERY ORGANISM OF EARTH, FROM BACTERIA TO HUMANS, SHARES A SINGLE UNIFYING SUBSTANCE THAT DICTATES EVERY ORGANISM'S GROWTH AND REPRODUCTION --

-- A DOUBLE-HELIX **SUPER-MOLECULE** EARTHLINGS CALL...

DNA!

D and...?

NO, D-N-A.--

IT'S AN ACRONYM FOR DEOXYRIBONUCLEIC ACID...

...WHICH, ALONG WITH RIBONUCLEIC ACID--

Whoa-- those are fancy words, Bloort.

Aren't we getting a little ahead of ourselves?

YOU'RE RIGHT, SIRE-- I'M JUST TRYING TO INTRODUCE SOME CONCEPTS HERE.

RIGHT NOW ALL I WANT YOU TO REMEMBER IS THAT DNA IS A MACROMOLECULE THAT EXISTS IN EVERY SINGLE LIVING CELL OF EVERY ORGANISM ON EARTH...

...AND THAT IT PROVIDES A HISTORY AND AN INSTRUCTION MANUAL FOR ANY GIVEN ORGANISM WRITTEN IN A UNIVERSAL LANGUAGE UNDERSTOOD BY EVERY CELL WITHIN THAT ORGANISM...

...A BLUEPRINT FOR MAINTAINING, AS WELL AS FOR BUILDING, CELLS...

TTATCG

TCCACATG

AGAG

ACGTTC GGTCAT

ATTCTA AAATGG

GCTA AGGG

ACCGTACC GTCCTCAT

TCCA

...AND FOR PASSING THE TRAITS OF THOSE CELLS ON TO FUTURE GENERATIONS.

Blueprints-- construction! I can understand **that**-- I've had some experience in **building**...

...an EMPIRE, not to brag.

But it takes more than blueprints and universal language-- or **whatever**-- to get something done.

It takes a BUREAUCRACY.

WELL, IT TAKES **ORGANIZATION**, AND DNA CERTAINLY **DOES** HAVE AN ORGANIZED METHOD.

A STRAND OF DNA IS COMPRISED OF THOUSANDS OF MOLECULES OF **NUCLEIC ACIDS**.

SPECIFIC **GROUPINGS** OF THOSE NUCLEIC ACIDS, WHICH ENCODE A **SPECIFIC PHYSICAL CHARACTERISTIC**, ARE CALLED **GENES**.

GENES ARE UNITS OF INFORMATION.

THE STRANDS OF DNA-- CONTAINING ALL THE GENETIC INFORMATION--

--ARE NEATLY PACKED INTO STRUCTURES KEPT WITHIN THE CELL CALLED **CHROMOSOMES**.

A COMPLETE SET OF AN ORGANISM'S GENES-- THE TOTALITY OF ALL THE INFORMATION NEEDED TO CREATE ANOTHER IDENTICAL ORGANISM --

--IS CALLED A **GENOME**.

EARTHLY ORGANISMS CAN HAVE ANYWHERE FROM **HUNDREDS** TO **TENS OF THOUSANDS** OF GENES WITHIN THEIR GENOME, REGARDLESS OF STRUCTURAL COMPLEXITY!

ROUNDWORM: 20,000 GENES

CHICKEN: 23,000 GENES

MAN: 20-25,000 GENES

AS EVIDENCE OF THEIR COMMON ORIGIN, GENES ARE REMARKABLY SIMILAR ACROSS ALL EARTHLY ORGANISMS. FOR INSTANCE, 31% OF THE GENES IN THAT BRAINY, GLOBALLY DOMINANT MAMMAL, THE **HUMAN**, AND IN THE SIMPLE, UNICELLULAR FUNGUS **BREWER'S YEAST** ARE EQUIVALENT...

...40% OF THE GENES IN THE HUMAN AND THE **ROUNDWORM** ARE EQUIVALENT...

...50% OF THE GENES IN THE HUMAN AND THE **FRUIT FLY** ARE EQUIVALENT...

...THE HUMAN AND THE **HOUSE MOUSE**, A WHOPPING 85%...

AND HERE'S SOMETHING INTERESTING -- HUMANS AND THEIR CLOSEST RELATIVE, THE **CHIMPANZEE**, SHARE MORE THAN **98%** OF THEIR GENES.

THEIR DIFFERENCES ARE ALMOST **NEGLIGIBLE**.

ALMOST negligible?

Personally, *I* can't see any difference at all.

LET US NOT STRAY INTO **SPECIES INSENSITIVITY,** YOUR MOST OBDURATE ALL-KNOWINGNESS.

WE SQUINCH, DON'T FORGET, HAVE MORE THAN A BIT IN COMMON WITH THE LOWLY **FRIPPURVIAN FLOGWORM...**

We do...?

Well, I suppose *THAT* could explain our genetic weakness...

UMM. I SERIOUSLY DOUBT IT, SIRE...

BUT, TO GET BACK TO EARTH...

...THE INCREDIBLE DIVERSITY OF LIFE WE SEE THERE, THE INCREDIBLE **ADAPTABILITY,** THE CAPACITY TO EXPERIMENT AND IMPROVE AND EVOLVE NEW SPECIES WITH RELATIVE SPEED...

...IS GREATLY FACILITATED THROUGH A REPRODUCTION STRATEGY THAT ALLOWS FOR THE SHARING OF GENETIC INFORMATION **BETWEEN** INDIVIDUALS WITHIN SPECIES.

IT IS AN **INSPIRED** STRATEGY CALLED...

SEX!

SHARING genetic information? Ugh.

That seems like it might be--unpleasant. INTRUSIVE, to say the least.

WELL, THERE **ARE** SOME DISADVANTAGES. AND LIFE ON EARTH **DID** BEGIN BY REPRODUCING JUST AS WE DO-- **ASEXUALLY**, AS IT'S CALLED THERE.

A LOT OF LOWER ORGANISMS THERE **STILL** PASS THEIR GENETIC HERITAGE TO THE NEXT GENERATION BY ESSENTIALLY **DUPLICATING** THEIR PERSONAL GENETIC MAKEUP--

--A PROCESS REFERRED TO ON EARTH AS **CLONING.**

Well, sure!

Like little JUNIOR here...

...Floorsh number 736 -- may the disorder not strike him!

OH, NEVER, YOUR MOST PRODUCTIVE FECUNDITY!

...BUT MOST OF THE MORE COMPLEX ORGANISMS ON EARTH HAVE CREATED A **DUALITY** WITHIN THEIR SPECIES. TWO VARIANTS OF A SPECIES HAVE EVOLVED -- TWO **SEXES**--CALLED **MALE** AND **FEMALE**.

THE MALE AND THE FEMALE MUST **COMBINE** THEIR GENETIC MATERIAL TO SUCCESSFULLY REPRODUCE.

THIS GUARANTEES THAT THE OFFSPRING WILL BE A REPLICATE OF **NEITHER** OF THE TWO PARENTS, BUT A COMPLETELY **UNIQUE** COMBINATION OF VARIOUS TRAITS INHERITED FROM THE TWO!

EVERY OFFSPRING OF SEXUAL REPRODUCTION IS SOME-THING NEW -- AN EXPERIMENT IN **INDEPENDENTLY AND RANDOMLY** RESHUFFLING THE PARENT GENES!

14

Seems needlessly complex.

And who would want a child who isn't exactly as expected?

A SPECIES UNDER PRESSURE IN A VERY COMPETITIVE ENVIRONMENT, THAT'S WHO!

IF YOUR SPECIES IS BEING OUT-COMPETED BY ANOTHER SPECIES FOR--SAY--A LIMITED FOOD SOURCE, YOU'D BETTER **HOPE** YOUR OFFSPRING GROW STRONGER, OR MORE RESOURCEFUL, THAN THE COMPETITION!

SEXUAL REPRODUCTION MARKEDLY SPEEDS UP THE RATE OF CHANGE WITHIN A SPECIES BECAUSE IT OFFERS MANY MORE VARIABLES, IN THE FORM OF GENES DESCENDING FROM **TWO** LINEAGES WITH ALL **THEIR** VARIATIONS.

ULTIMATELY, PROBABILITY DICTATES THAT SOME OF THOSE INFINITELY VARIABLE COMBINATIONS OF PARENTAL TRAITS WILL PRODUCE POSITIVE CHANGES-- **COMPETITIVE ADVANTAGES** FOR THE SPECIES.

OF COURSE, IF THE COMPETITION IS REPRODUCING SEXUALLY, TOO, THEN THEY'RE PROBABLY KEEPING UP WITH YOU ON THE EVOLUTIONARY BATTLEFIELD.

THAT'S THE DOWNSIDE OF SEX-- IT TAKES AN AWFUL LOT OF ENERGY, AND THERE IS A LOT OF WASTE. MOST NEW COMBINATIONS OF GENES DON'T OFFER ANY USEFUL ADVANTAGE, AND SOME COMBINATIONS ARE HURTFUL...

...BUT ON THE VERY COMPETITIVE PLANET EARTH, OVER THE LONG HAUL, IT HAS PROVEN TO BE THE PROGRESSIVE, WINNING STRATEGY.

SPLORT!!

OUCH.

ASEXUAL REPRODUCTION IS SAFE-- YOU KNOW WHAT YOU'RE GOING TO GET-- BUT THAT CONSISTENCY LEADS TO A COMPARATIVELY **SLOW** RATE OF EVOLUTION.

PRIMATES EMPLOY **SEXUAL REPRODUCTION** --AND IN **HUMANS**, THEY HAVE DEVELOPED A WORLD-DOMINATING INTELLIGENCE...

...IN LESS THAN 5 MILLION YEARS!

Less than five...

That **is** fast, isn't it?

IT TOOK US A LOT LONGER TO GROW OUR BRAINS.

Okay, I think I see where this is going...

You **REALLY** believe something of this sex business can help us?

YES, I DO, YOUR MOST INCISIVE PERCEPTIVENESS.

I WAS SENT TO EARTH WITH A SPECIFIC MISSION-- TO LEARN HOW THAT PLANET'S LIFE DEALS WITH INHERITED DISEASE...

I am **WELL** aware of the mission, Bloort 183! The genetic disorder afflicting my household is something I must deal with **EVERY DAY.**

The rest of Squinchdom, too-- I feel for **ALL** my subjects.

And I don't like the word DISEASE.

Y-YES, SIRE...

THE FACT IS THAT WHILE OUR CONSERVATIVE METHOD OF SELF-REPRODUCING HAS KEPT OUR HEALTH REMARKABLY STABLE FOR MOST OF OUR EXISTENCE, IT HAS ALSO INHIBITED OUR ABILITY TO RESPOND TO NEW CHALLENGES...

...AS WE NOW SEE WITH THE GROWTH OF THESE AWFUL GENETIC ...DISORDERS, AGAINST WHICH WE HAVE NO DEFENSE.

WE HAD HOPED THAT MY STUDY OF EARTH LIFE MIGHT OFFER A SOLUTION-- AND I THINK IT HAS!

I'M CONVINCED THAT THE ADVANTAGES THAT COME WITH EXCHANGING GENES WILL BE USEFUL IN STRENGTHENING SQUINCH BIOLOGY!

IT IS NOT NECESSARY TO HAVE TWO DIFFERING SETS OF REPRODUCTIVE ORGANS. SEXUAL REPRODUCTION CAN BE ACHIEVED ARTIFICIALLY, AS YOU WILL SEE.

THE HUMAN ORGANISMS HAVE LEARNED **AMAZING** THINGS ABOUT THEIR GENETIC LEGACY AND ARE ON THE VERGE OF CREATING A REVOLUTION IN GENETIC HEALTH FOR THEMSELVES.

THERE IS MUCH WE CAN BORROW FROM THEM.

THEY CLEARLY HAVE HIGH EXPECTATIONS FOR THEM-SELVES...

IN THEIR SCIENTIFIC LANGUAGE, HUMANS DESIGNATE THEMSELVES AS *HOMO SAPIENS*, WHICH MEANS "WISE, KNOWING MAN."

ALTHOUGH, IN A SIMILAR PARLANCE, **I** WOULD CHOOSE TO CALL THEM **SOPHOMORES** --

"WISE FOOLS," ONLY PARTWAY THROUGH THEIR SELF-EDUCATION, IN CONSTANT DANGER OF MAKING TERRIBLE MISTAKES.

THEY WRESTLE CONTINUALLY WITH THE **ETHICAL** AND **PRACTICAL** DILEMMAS INHERENT IN THEIR EVERY DISCOVERY AND INVENTION...

...WHICH FOLLOW ONE UPON THE OTHER WITH **TREMENDOUS** SPEED.

Because of -- **SEX**?

NOT EXACTLY, SIRE...

ACTUALLY-- YES, YOUR GIGANTIC LORDLINESS!

SEX DOES HAVE A LOT TO DO WITH IT!

IT WAS **SEXUAL REPRODUCTION** THAT FIRST LED MAN TO BEGIN HIS UNDERSTANDING OF THE POSSIBILITIES OF GENETICS...

SEVERAL THOUSAND YEARS AGO, A HUMAN -- MAYBE A KEEPER OF DOMESTICATED ORGANISMS -- FIRST RECOGNIZED A BASIC, BUT IMPORTANT, PRINCIPLE: **LIKE BEGETS LIKE.**

SHEEP ALWAYS PRODUCE SHEEP, AND GRAPES, MORE GRAPES. GRAPES **NEVER** PRODUCE SHEEP.

THE HUMANS' GRASP OF THE TRUISM THAT **ORGANISMS REPRODUCE THEMSELVES** LED THEM TO OBSERVE THAT SPECIFIC TRAITS COULD BE PASSED FROM PARENT TO OFFSPRING.

THIS LED TO THE **SELECTIVE BREEDING** OF ANIMALS AND PLANTS WITH CHARACTERISTICS THEY FOUND DESIRABLE...

...AND THE DELIBERATE EXCLUSION OF THOSE WITH UNDESIRABLE TRAITS.

OBSERVATION SHOWED WHERE HUMANS WERE RIGHT AND WHERE THEY WERE WRONG.

NATURE CHANGES SPECIES BLINDLY, OVER LONG PERIODS OF TIME, BY PURE TRIAL AND ERROR. HUMANS CONSCIOUSLY MANIPULATED GENES WITH A CRITICAL, PURPOSE-DRIVEN EYE AND SAW RESULTS WITH REWARDING SPEED.

THIS WAS THE **BEGINNING** OF THE UNDERSTANDING OF **GENETICS**.

BUT THAT WAS PRETTY MUCH **IT** FOR THOUSANDS OF YEARS. ONGOING ATTEMPTS TO GRASP THE **UNDERLYING MECHANICS** OF REPRODUCTION SPROUTED SOME INTERESTING THEORIES-- **SPONTANEOUS GENERATION** WAS POPULARLY ACCEPTED FOR A LONG TIME--

Behold! Dirty sheets give rise to bedbugs!

--BUT-THEY WERE ALMOST ALWAYS STEPS **BACKWARD.**

IT WAS LESS THAN 200 YEARS AGO THAT TWO PARTICULAR HOMO SAPIENS--HUMANS, THAT IS-- FINALLY STARTED THINGS ON THE RIGHT TRACK.

CHARLES DARWIN and **GREGOR MENDEL**

ACCURATELY CONCLUDED THAT SPECIES **EVOLVE** BASED ON THE ABILITIES OF INDIVIDUAL ORGANISMS TO **SURVIVE** AND **REPRODUCE.**

ACCURATELY OBSERVED AND DESCRIBED **HOW** TRAITS ARE PASSED FROM ONE GENERATION TO THE NEXT.

USING THE **SCIENTIFIC METHOD** AND **CONTROLLED EXPERIMENTS**, MENDEL OBSERVED GENERATIONS OF CAREFULLY CULTIVATED PEA PLANTS AND FROM THEM DETERMINED THE **LAWS OF INHERITANCE.**

HE FOUNDED THE **SCIENCE** OF GENETICS, ALTHOUGH IT WOULD BE DECADES BEFORE HIS BREAKTHROUGHS WERE PROPERLY RECOGNIZED.

BUT ONCE SCIENCE **DID** ACCEPT MENDEL'S GENES, THE MYSTERIES OF THE MECHANICS BEHIND HEREDITY WERE SOLVED WITH REMARKABLE RAPIDITY.

JAMES WATSON

FRANCIS CRICK

THE RELATIONSHIP BETWEEN GENES AND CHROMOSOMES WAS DISCOVERED -- DNA WAS **DESCRIBED** -- AND THE COMPOSITION AND STRUCTURE OF DNA WAS **UNRAVELED** ...

... AND NOW THE GENETIC LANGUAGE ITSELF IS BEING TRANSLATED AND INTERPRETED!

HUMANITY IS LEARNING TO **READ** THE BOOK OF LIFE!

PERHAPS THE CROWNING ACHIEVEMENT HAS BEEN THE **HUMAN GENOME PROJECT** --

CHEMISTRY BIOLOGY PHYSICS ETHICS ENGINEERING INFORMATICS

-- A MASSIVE EFFORT THAT HAS MAPPED THE **ENTIRETY** OF HUMAN DNA -- ALL **3 BILLION BASE PAIRS** (I'LL EXPLAIN LATER).

All right -- so they've gotten to know themselves intimately.

How does this help us deal with our problems?

THE KNOWLEDGE THEY'VE GAINED **ISN'T** JUST FOR ITS OWN SAKE.

THEY'RE **APPLYING** WHAT THEY'VE LEARNED.

THEY'RE CHANGING THEIR WORLD -- AND **THEMSELVES** -- THROUGH THEIR GROWING UNDERSTANDING OF GENETICS.

MAYBE MOST IMPORTANT, THEY ARE LEARNING HOW TO ALLEVIATE THE SUFFERING THAT COMES WHEN DEFECTIVE OR BADLY MATCHED GENES CAUSE DISORDERS AND DISEASES, OR WHEN CHROMOSOMES ARE DAMAGED.

BASIC KNOWLEDGE OF HEREDITY HAS TAUGHT THEM TO TRACE THEIR ANCESTRIES. AN UNDERSTANDING OF THE SPECIFIC GENETIC TRAITS IN A FAMILY HISTORY CAN HELP INDIVIDUALS ESTIMATE THEIR CHANCES OF PASSING ON THOSE TRAITS TO THEIR CHILDREN -- QUITE IMPORTANT WHEN IT COMES TO **AVOIDING** HEREDITARY GENETIC DISORDERS.

IT MIGHT BE HELPFUL IF **WE** COULD LEARN TO CALCULATE THE DANGERS INHERENT IN **OUR** LINEAGES!

SOME GENETIC PROBLEMS POP UP **WITHOUT** A HISTORY. GENETIC CHANGE -- **MUTATION** -- OCCURS WITHIN INDIVIDUALS, AND CAN BE EITHER GOOD OR BAD.

CANCERS ARE DISEASES OF DNA THAT HAS BEEN DAMAGED BY MUTATION.

NORMAL

CANCEROUS

HUMANITY IS USING ITS KNOWLEDGE OF DNA TO CONSTRUCT THERAPIES FOR MANY OF THESE GENETIC PROBLEMS.

ONE OF THE MOST PROMISING THERAPIES INVOLVES CONVERTING **VIRUSES** TO DELIVERY **VECTORS** CARRYING HEALTHY GENES TO AFFLICTED CELLS.

VIRUSES NATURALLY **HIJACK** CELLS TO REPRODUCE THEMSELVES.

HEALTHY GENES

HUMAN GENETICISTS HAVE LEARNED TO TAME THESE TINY CAPSULES, LOAD THEM WITH HEALTHY GENES ...

... AND LAUNCH THEM AT THE CELLS OF THE APPROPRIATE TISSUE. THE VIRUSES, DOING WHAT COMES NATURALLY, DELIVER THEIR PAYLOAD.

Le' me at 'im!

ZOOM

BUT INSTEAD OF REPRODUCING THEMSELVES, THEY INTRODUCE NEW, HEALTHY HUMAN GENES TO REPLACE THE DAMAGED ONES. **AMAZING!** THIS IS CALLED **RECOMBINANT DNA TECHNOLOGY.**

THIS TECHNOLOGY IS USED FOR OTHER PURPOSES AS WELL. HUMAN DNA IS BEING RECOMBINED WITH **BACTERIA** -- WHICH REPRODUCE QUICKLY AND UNDER HIGHLY CONTROLLED CONDITIONS -- TO PRODUCE LARGE AND AFFORDABLE QUANTITIES OF PREVIOUSLY DIFFICULT-TO-MASS-PRODUCE DRUGS SUCH AS **INSULIN** AND **INTERFERON**, WHICH ARE USED TO COUNTER THE EFFECTS OF DAMAGED GENES.

eep!

E. COLI

THE TECHNIQUE OF RECOMBINANT DNA HAS EVEN BEEN HARNESSED TO CREATE CERTAIN **PLANTS** THAT PRODUCE HUMAN **ANTIBODIES!** THESE ARE EASILY HARVESTED AND USED, ONCE AGAIN, TO FIGHT DISEASES.

?!

WHAT NEXT?

HUMANITY HAS LONG UNDERSTOOD THE PRINCIPLE OF **CLONING**, AND HAS EMPLOYED IT FOR CENTURIES AS A TOOL IN KEEPING CONSISTENT QUALITIES IN AGRICULTURAL PRODUCE, SUCH AS THE FRUIT CALLED **APPLES**.

BUT IT WASN'T UNTIL RECENTLY THAT THEIR SKILLS ALLOWED THEM TO BEGIN CLONING **ANIMALS**. ALTHOUGH THEY'VE HAD LIMITED SUCCESS UP UNTIL NOW, IT IS PROBABLE THAT THEY WILL SOONER OR LATER PERFECT THIS TECHNIQUE.

AND LET ME TELL YOU, THERE ARE SOME HEATED DISCUSSIONS ABOUT **THIS** BREWING ON EARTH!

SOME HUMANS HAIL CLONING AS AN ANSWER TO THEIR LIMITED LIFE SPANS -- A POSSIBILITY FOR INDIVIDUALS TO KEEP BANKS OF THEIR OWN CLONED ORGANS AS REPLACEMENTS FOR THE WORN-OUT ORIGINALS.

BOB'S XTRA HEART

BOB'S XTRA KIDNEY

BOB'S XTRA LIVER

BOB'S XTRA TESTES

BOB'S XTRA LUNG

BOB'S XTRA SPLEEN

BOB

SOME EVEN BELIEVE THAT **DNA ITSELF** MAY BE CLONED TO INDEFINITELY DELAY THEIR AGING PROCESS!

AS YOU MIGHT IMAGINE, THIS IS RAISING PLENTY OF ETHICAL QUESTIONS -- AS DO CERTAIN HUMANS' DREAMS OF PURPOSE-FULLY CHOOSING TO **GENETICALLY ENGINEER** DESIRED TRAITS INTO FUTURE GENERATIONS.

...BABIES EXPRESSLY DESIGNED TO ORDER...

...BELOVED PETS CLONED TO LAST FOREVER.

BUT THE HUMANS' EFFORTS TO MASTER DNA ARE CLEARING UP DILEMMAS AS WELL.

CRIMES ARE BEING SOLVED AND LEGAL ISSUES UNTANGLED WITH THE USE OF **DNA PROFILING**, WHERE PHYSICAL EVI-DENCE IS ANALYZED FOR THE **GENETIC "FINGERPRINTS"** OF INDIVIDUALS.

FINALLY, THE GROWING UNDER-STANDING OF THEIR GENOME HAS UNLOCKED MANY OF THE **MYSTERIES** OF THEIR PREHISTORIC PAST, ALLOWING THEM A GLIMPSE INTO THEIR ORIGINS ON THE CONTINENT CALLED **AFRICA** AS WELL AS THEIR SUCCESSFUL DISPERSION AROUND THE GLOBE.

AS THEY LEARN ABOUT THE GENETIC STRENGTHS AND WEAKNESSES OF THEIR GEOGRAPHICALLY DIVERSE POPULATIONS...

AFRICA

ARABIA

... THEY GAIN A DEEPER UNDERSTANDING OF THEMSELVES -- THEIR ABILITIES, THEIR LIMITS, THEIR POTENTIAL.

THEY ARE A SPECIES CONSTANTLY PUSHING TO IMPROVE THEMSELVES AND THEIR WORLD...

... AND THEY STRUGGLE EVERY **INCH** OF THE WAY WITH ALL THE COMPLEXITIES THAT ENTAILS.

THE RELATIVE VALUES OF SHORT-RANGE GAINS IN HEALTH WEIGHED AGAINST LONG-RANGE GENETIC DANGERS--

-- IMMEDIATE PROFIT WEIGHED AGAINST UNIMAGINABLE BACKLASHES IN THE DISTANT FUTURE...

...WHEW!

LIFE ON EARTH AND THE DNA THAT MAKES IT WORK ARE REMARK-ABLE THINGS.

HUMANITY IS A REMARKABLE SPECIES, AND I REPEAT-- WE CAN LEARN FROM THEM. ENOUGH, I THINK, TO HELP OURSELVES.

NOW THAT I'VE CONCLUDED MY **INTRODUCTION**, YOUR MOST DISCRIMINATING LORDLINESS, DO I HAVE YOUR PER-MISSION TO CONTINUE?

Of **COURSE**! I've heard enough to be intrigued.

Get **ON** with it, for pity's sake...

We have much to learn, and little time.

VERY GOOD, THEN, SIRE.

THERE ARE MANY PLACES WE COULD START, BUT I SUGGEST WE BEGIN WITH THE **BASICS**, WITH THE ELEMENTS THAT COMPRISE US **ALL**...

...NO MATTER WHAT **WORLD** WE COME FROM.

CHAPTER 1

How the System Works--The Molecular Story

WE'LL START WITH A **CELL**-- THE BASIC BUILDING UNIT OF ALL EARTHLY ORGANISMS.

FOR OUR PURPOSES, A **EUKARYOTE** CELL-- THAT IS, ONE WITH A **NUCLEUS.** LET'S SAY IT'S A CELL FROM A **HUMAN.**

AS WE WANT TO UNDERSTAND THE MOST BASIC MECHANICS GUIDING LIFE'S PROCESSES, WE NEED TO LOOK EVEN **CLOSER** -- INTO THE CELL'S **NUCLEUS,** THE ORGANELLE THAT PROTECTS THE **CHROMOSOMES** AND THEIR COILED STRANDS OF **DNA**...

...AND THEN INTO THE DNA ITSELF AND RIGHT DOWN TO THE MOST BASIC ELEMENTS THAT COMPRISE ALL MATTER THROUGHOUT THE UNIVERSE--

--**ATOMS!**

 OXYGEN
O

 PHOSPHORUS
P

CARBON
C

SULFUR
S

HYDROGEN
H

NITROGEN
N

ESSENTIAL TO LIFE ON EARTH ARE **SIX KEY ATOMIC ELEMENTS**--INCLUDING **CARBON,** THAT VERY USEFUL COMPONENT INTRODUCED TO THIS WORLD BY METEORITES--

--ALL OF WHICH ARE VERY GOOD **JOINERS.**

PHOSPHATE (PO_4)

WATER (H_2O)

THEY WORK WELL TOGETHER, LINKING UP IN VARIOUS COMBINATIONS TO FORM SIMPLE BUT IMPORTANT **MOLECULES** LIKE **WATER** AND **PHOSPHATE** ...

... AS WELL AS THE MORE COMPLEX **CARBOHYDRATE** MOLECULES CALLED **SUGARS**, WITH THEIR DISTINCTIVE RING-SHAPED STRUCTURES.

GLUCOSE ($C_6H_{12}O_6$)

THESE RELATIVELY SIMPLE MOLECULES ARE GOOD AT WORKING TOGETHER, TOO. THEY CAN FORM **MACROMOLECULES** THAT CHAIN TOGETHER **HUNDREDS**, OR EVEN **THOUSANDS**, OF SIMPLER MOLECULES, LIKE A STRING OF BEADS.

POLYSACCHARIDES -- LIKE STARCH OR CELLULOSE -- ARE SIMPLY CHAINS OF **SUGAR** MOLECULES...

... WHILE **LIPIDS** ARE CHAINS OF **HYDROCARBONS** THAT **REPEL WATER** AND ARE THUS VERY USEFUL IN FORMING PROTECTIVE MEMBRANES.

AMONG THE **MOST** COMPLEX MACROMOLECULES ARE **THE PROTEINS.** THESE ARE CHAINS OF **AMINO ACIDS** RESPONSIBLE FOR **MANY** METABOLIC, STRUCTURAL, AND MECHANICAL PROCESSES.

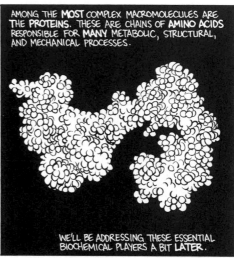

WE'LL BE ADDRESSING THESE ESSENTIAL BIOCHEMICAL PLAYERS A BIT **LATER.**

FIRST WE NEED TO LOOK AT THE FAMILY OF MACROMOLECULES RESPONSIBLE FOR THE **DESIGN** AND **ASSEMBLY** OF THOSE ALL-IMPORTANT PROTEINS -- THE **NUCLEIC ACIDS**...

... DEOXYRIBONUCLEIC ACID-- OR **DNA**--

... AND RIBONUCLEIC ACID-- OR **RNA**!

THE NUCLEIC ACIDS ARE MADE UP OF COMPONENTS CALLED **NUCLEOTIDES**. EACH NUCLEOTIDE IS COMPOSED OF ONE **PHOSPHATE** MOLECULE, ONE **SUGAR** MOLECULE, AND A NITROGEN-RICH **BASE**-- ALL OF WHICH ARE SECURELY BONDED TOGETHER.

NUCLEOTIDE

THE **SUGAR** MOLECULE OF A NUCLEOTIDE CAN COME IN ONE OF TWO VARIETIES--

RIBOSE

--EITHER A **RIBOSE**...

...OR A **DEOXYRIBOSE**. YOU'LL NOTICE THAT DEOXY-RIBOSE IS MISSING THE **OXYGEN ATOM** THAT RIBOSE CARRIES DOWN THERE IN ITS BOTTOM RIGHT-HAND CORNER. AND-- YOU GUESSED IT-- THE PRESENCE OR ABSENCE OF THAT ONE OXYGEN ATOM IS THE DIFFERENCE BETWEEN RNA AND DNA.

DEOXYRIBOSE

THE **BASE** MOLECULE IN A NUCLEOTIDE COMES IN ONE OF FOUR DIFFERENT TYPES: **ADENINE (A)**, **THYMINE (T)**, **CYTOSINE (C)**, OR **GUANINE (G)**.

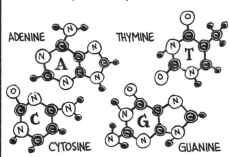

ADENINE   THYMINE

CYTOSINE   GUANINE

IN RNA, **URACIL (U)** SUBSTITUTES FOR THYMINE, WHILE THE OTHER BASES REMAIN THE SAME. SEEMS NEEDLESSLY COMPLEX, I KNOW, BUT THERE IS A GOOD REASON AND WE'LL ADDRESS THAT **LATER**. BUT, RIGHT NOW, LET'S FOCUS ON **DNA**.

NOTICE HOW ADENINE AND GUANINE BOTH ARE BUILT WITH **DOUBLE-RING STRUCTURES**. CYTOSINE AND THYMINE, MEANWHILE, HAVE **SINGLE-RING STRUCTURES**. LIKE PREDESTINED LOVERS, THE A IS ATTRACTED TO THE T AND FORMS A STRONG ATOMIC BOND. LIKEWISE, G AND C ARE BUILT TO BOND!

SO, IN DNA, A STRONG BOND IS CREATED **WITHIN** NUCLEOTIDES AS PAIRS OF COMPLEMENTARY BASES HOOK UP. AT THE SAME TIME, SUGARS AND PHOSPHATES BOND, WHICH CAUSES THE NUCLEOTIDES THEMSELVES TO HOOK UP. ALL OF WHICH MAKES FOR A VERY STABLE, VERY DURABLE MACROMOLECULE!

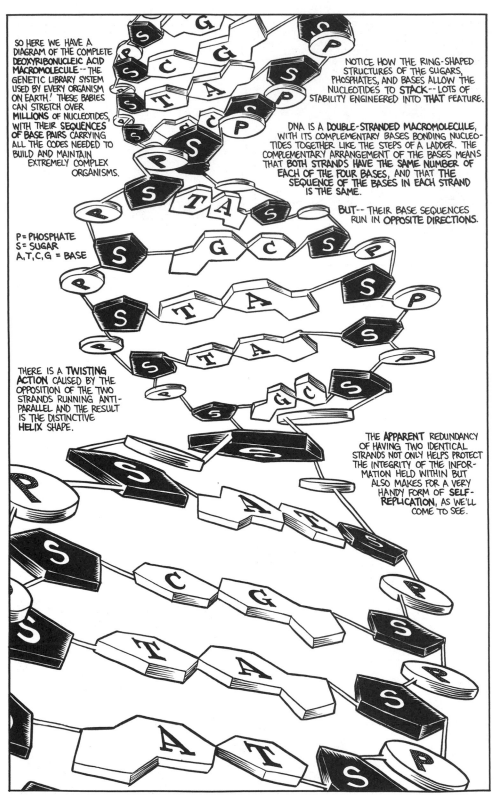

SO HERE WE HAVE A DIAGRAM OF THE COMPLETE **DEOXYRIBONUCLEIC ACID MACROMOLECULE** -- THE GENETIC LIBRARY SYSTEM USED BY EVERY ORGANISM ON EARTH! THESE BABIES CAN STRETCH OVER **MILLIONS** OF NUCLEOTIDES, WITH THEIR **SEQUENCES OF BASE PAIRS** CARRYING ALL THE CODES NEEDED TO BUILD AND MAINTAIN EXTREMELY COMPLEX ORGANISMS.

NOTICE HOW THE RING-SHAPED STRUCTURES OF THE SUGARS, PHOSPHATES, AND BASES ALLOW THE NUCLEOTIDES TO **STACK** -- LOTS OF STABILITY ENGINEERED INTO **THAT** FEATURE.

DNA IS A **DOUBLE-STRANDED MACROMOLECULE,** WITH ITS COMPLEMENTARY BASES BONDING NUCLEOTIDES TOGETHER LIKE THE STEPS OF A LADDER. THE COMPLEMENTARY ARRANGEMENT OF THE BASES MEANS THAT BOTH STRANDS HAVE THE SAME NUMBER OF EACH OF THE FOUR BASES, AND THAT THE SEQUENCE OF THE BASES IN EACH STRAND IS THE SAME.

BUT-- THEIR BASE SEQUENCES RUN IN **OPPOSITE DIRECTIONS.**

P = PHOSPHATE
S = SUGAR
A, T, C, G = BASE

THERE IS A **TWISTING ACTION** CAUSED BY THE OPPOSITION OF THE TWO STRANDS RUNNING ANTI-PARALLEL AND THE RESULT IS THE DISTINCTIVE **HELIX SHAPE.**

THE **APPARENT** REDUNDANCY OF HAVING TWO IDENTICAL STRANDS NOT ONLY HELPS PROTECT THE INTEGRITY OF THE INFORMATION HELD WITHIN BUT ALSO MAKES FOR A VERY HANDY FORM OF **SELF-REPLICATION,** AS WE'LL COME TO SEE.

EACH INCREDIBLY LONG STRAND OF DNA IS DIVIDED INTO DISCRETE COHESIVE **GROUPS** OF NUCLEOTIDES. THESE ARE THE **GENES**-- EACH WITH A SPECIFIC SEQUENCE OF BASES THAT CODES FOR **ONE SPECIFIC PROTEIN.** PHYSICAL TRAITS--ANYTHING FROM EYE COLOR TO HEIGHT TO THE NUMBER OF LIMBS-- ARE MANIFESTED THROUGH THE INTERACTION OF EXPRESSED GENES (PROTEINS).

ONE GENE

ONE PROTEIN

DNA MACROMOLECULES ARE SO LONG THAT THEY WOULD NEVER FIT INSIDE THE CELL'S NUCLEUS WITHOUT A SPACE-SAVING TECHNIQUE CALLED **SUPERCOILING.** SEGMENTS OF THE DNA WRAP AROUND AND AROUND A COMPLEX OF PROTEINS CALLED **HISTONES**--LIKE A LONG THREAD WOUND ONTO HOLLOW SPOOLS.

THIS SECURES THE DNA AND PREVENTS IT FROM UNRAVELING. A DNA SEGMENT WRAPPED AROUND HISTONES IS CALLED A **NUCLEOSOME.**

HISTONE

IN THIS MANNER, HUMAN DNA STORES ITS **THREE BILLION BASE PAIRS** INSIDE A MICROSCOPIC PINPOINT!

ACTUALLY, THE DNA MACROMOLECULE DOES NOT STORE AS ONE, UNINTERRUPTED STRAND. PACKAGED AS NUCLEOSOMES, THE DNA IS BROKEN UP INTO **RELATIVELY** SHORT UNITS CALLED **CHROMOSOMES.**

THE NUMBER OF CHROMOSOMES VARIES AMONG ORGANISMS, BUT IN HUMANS A FULL SET EQUALS 46 AND--

WHOA!

Maybe I'm just slow on the uptake, but...

...I'm having trouble keeping everything straight!

THAT'S NOT SURPRISING, YOUR ROYAL INSIGHTFULNESS!

THE LANGUAGE OF GENETIC SCIENCE IS VERY **SPECIFIC** AND **SPECIALIZED**-- IT NEEDS TO CONVEY CONCEPTS THAT ARE SOMETIMES HARD TO VISUALIZE. IT TAKES SOME STUDY TO MAKE SENSE OF IT.

LET ME TRY TO EMPHASIZE THE MOST IMPORTANT POINTS I'VE PRESENTED WITH AN **ANALOGY...**

LET'S IMAGINE THAT THE MOST BASIC PHYSICAL UNITS, **ATOMS**, WHICH COMBINE INTO **MOLECULES**...

...ARE SIMPLE MACHINES THAT CAN BE BUILT INTO **COMPLEX MECHANICAL DEVICES**.

AND THAT A STRAND OF **DNA**, COMPOSED OF MANY, MANY MOLECULES...

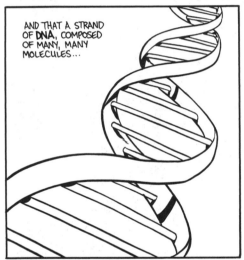

...IS A **COMPLEX MACHINE**, LIKE A **COMPUTER**, THAT STORES AN **ENORMOUS DATABASE**.

FOLLOWING THIS ANALOGY FURTHER, **GENES** WITHIN THE DATABASE WOULD BE SPECIFIC **DATA FILES** OF ARCHITECTURAL INFORMATION...

...AND **CHROMOSOMES** CAN BE VIEWED AS A SERIES OF **HARD DRIVES**, AMONG WHICH THE DATA IS STORED.

NOW CONSIDER THE **NUCLEUS** A KIND OF **SAFE** WITHIN WHICH THE HARD DRIVES ARE SECURED-- AND THERE YOU HAVE THE GENERAL IDEA!

NOW WE'RE ON THE PATH TO UNDERSTANDING HOW DNA TURNS ITS CODED INSTRUCTIONS INTO CHEMICAL PRODUCTION.

BUT BEFORE WE GET **THERE** WE NEED TO LEARN ABOUT DNA REPLICATION...

...THAT IS, HOW DNA DUPLICATES ITSELF WHEN A CELL DIVIDES TO CREATE MORE CELLS.

CELL DIVISION IS NECESSARY FOR ANY HEALTHY, GROWING ORGANISM-- WE'LL EXPLORE THE ENTIRE PROCESS IN MORE DETAIL LATER. RIGHT NOW WE'LL LOOK AT JUST THE **DNA**.

IT MUST DIVIDE WITH THE CELL, BECAUSE **EVERY** NEW CELL MUST CARRY A COMPLETE GENOME...

BY WHICH, OF COURSE, YOU MEAN A FULL COMPLEMENT OF THE ORGANISM'S GENES!

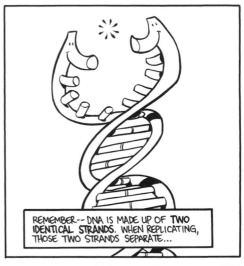

REMEMBER-- DNA IS MADE UP OF **TWO IDENTICAL STRANDS**. WHEN REPLICATING, THOSE TWO STRANDS SEPARATE...

...WITH EACH STRAND SERVING AS A **TEMPLATE** TO FORM A NEW, COMPLETE DNA MOLECULE! THIS SEMI-CONSERVATIVE PROCESS ALLOWS EVERY DNA MOLECULE TO BE MADE OF **ONE HALF** OF ITS "PARENT" AND HELPS KEEP REPLICATION MISTAKES TO A MINIMUM.

BOY, YOU LOOK FAMILIAR...

WHEN A CELL GETS SET TO DIVIDE, IT SENDS A **CHEMICAL MESSAGE** TO THE NUCLEUS IN THE FORM OF SPECIAL **ENZYMES**. THEIR MESSAGE IS TO TELL THE DNA THAT IT'S TIME TO REPLICATE...

WAKE UP! TIME TO REPLICATE!

THE MESSENGERS START THE PROCESS BY HELPING THE DNA SEGMENTS **UNWIND** FROM THEIR **NUCLEOSOME** FORM. THE DNA **MUST** BE UNWOUND FOR REPLICATION TO BEGIN.

IT'S ENOUGH TO MAKE A MOLECULE DIZZY!

INITIATOR

INITIATOR

THIS ALLOWS ANOTHER SET OF SPECIAL PROTEINS CALLED **INITIATORS** TO SCAN THE TWO STRANDS FOR CERTAIN SEQUENCES OF BASES CALLED **ORIGINS**.

THERE COULD BE **THOUSANDS** OF THESE **ORIGINS** DISTRIBUTED THROUGHOUT THE INCREDIBLY LONG DNA MOLECULE, AND THEY TELL THE INITIATORS THAT THE REPLICATION PROCESS MUST START **HERE** -- AT THESE POINTS. AN INDIVIDUAL REPLICATION PROCESS BEGINS WITH EVERY ORIGIN, PROCEEDING SIMULTANEOUSLY FROM EACH.

START HERE!

IT WOULD TAKE **FOREVER** TO DO THIS JOB WITHOUT A WHOLE LOTTA SIMULTANEOUS REPLICATION!

ONCE AN INITIATOR HAS DETERMINED AN ORIGIN, IT **FORCES AN OPENING** BETWEEN THE INDICATED COMPLEMENTARY BASES.

START HERE!

THIS ALLOWS ANOTHER PLAYER -- THE ENZYME **HELICASE** -- TO **REALLY** GET BUSINESS STARTED. HELICASE FINDS THE OPENINGS STARTED BY THE INITIATORS AND BEGINS **BREAKING APART** SEQUENCES OF THE DOUBLE HELIX HUNDREDS OF BASES LONG.

HELICASE

COMIN' THROUGH!

PROTEINS CALLED **SINGLE-STRANDED BINDERS** AID HELICASE'S WORK BY KEEPING THE BROKEN BASE SEQUENCES **PROPPED OPEN**, WHILE YET ANOTHER PROTEIN CALLED **GYRASE** PREVENTS THE UNWINDING HELIX FROM **TORQUING** AND **KNOTTING** LIKE A BAD TELEPHONE CORD.

GYRASE

S.S.B.

HELICASE

S.S.B.

As you can see here, Helicase's attack forms **TWO FORKS** in the molecule. Replication can proceed down **BOTH** of the two template strands in opposite directions.

NOW, TO GET THE ACTUAL REPLICATION ITSELF STARTED, A NUCLEOTIDE OF **TEMPORARY RNA** (RIBO-NUCLEIC ACID -- DNA's MORE INTERACTIVE COUSIN) COMES ALONG AND BONDS WITH THE BASE OF THE FIRST NUCLEOTIDE OPENED BY HELICASE.

HELLO THERE, SAILOR.

RNA NUCLEOTIDE

START HERE!

THIS BIT OF TEMPORARY RNA IS ESSENTIALLY **PRIMING THE PUMP** -- ONCE IT HAS GOT THE PROCESS ROLLING, ANOTHER ENZYME, **DNA POLYMERASE**, TAKES OVER AND DOES THE SERIOUS HEAVY LIFTING. DNA POLYMERASE MOVES DOWN THE OPENED CHAIN, IDENTIFIES THE BASES IN SEQUENCE, AND CONNECTS EACH WITH A **DNA NUCLEOTIDE** OF THE APPROPRIATE COMPLEMENTARY BASE.

THANK YOU, RNA-- I'LL TAKE IT FROM HERE.

DNA POLYMERASE

AND SO A COMPLETE, NEW DNA MOLECULE IS FORMED, IN ALL ITS DOUBLE-STRANDED GLORY, WITH **HELICASE** CONSTANTLY PLUNGING AHEAD, OPENING NEW BASES DOWN THE CHAIN...

...AND **POLYMERASE** FOLLOWING, ADDING COMPLEMENTARY NUCLEOTIDES, BASE BY BASE.

THIS PROCESS CONTINUES ON IN BOTH STRANDS OF THE TEMPLATE MOLECULE...

IF YOU THINK **THIS** IS COMPLEX, YOU SHOULD CHECK OUT **OKAZAKI FRAGMENTS!**

...UNTIL **TWO NEW DNA MOLECULES** ARE COMPLETE, EACH COMPRISED OF **ONE ORIGINAL PARENT STRAND** AND ONE **NEW STRAND** SYNTHESIZED FROM FREE-FLOATING NUCLEOTIDES.

ACTUALLY, WE'RE **BOTH** STILL A BIT FRAGMENTED, THANKS TO OUR MULTIPLE ORIGINS...

THERE ARE STILL SOME FINISHING TOUCHES THAT NEED TO BE ATTENDED TO. THOSE RNA PRIMER NUCLEOTIDES THAT GOT REPLICATION STARTED MUST BE REMOVED BY **DNA POLYMERASE**, AND REPLACED WITH ACTUAL **DNA** NUCLEOTIDES HAVING THE APPROPRIATE BASES...

...AND THEN ALL THE FRAGMENTED SECTIONS OF THE NEW DNA MOLECULE -- EACH STARTED AT ITS OWN ORIGIN -- MUST BE JOINED TOGETHER BY THE ENZYME **LIGASE**, LINKING PHOSPHATES TO SUGARS WITH THEIR POWERFUL BOND.

THAT'S A **LOT** OF ENZYMES AND A **LOT** OF STEPS TO COMPLETE, AND REPLICATION TRANSPIRES **QUICKLY**. IN HUMANS, 2,000 BASES ARE REPLICATED PER MINUTE! GIVEN THE HIGH DEGREE OF DIFFICULTY, IT'S NO WONDER THAT MISTAKES -- ALTHOUGH AMAZINGLY INFREQUENT -- DO OCCUR.

BASES **DO** OCCASIONALLY GET MISMATCHED. THE RESULT IS A BAD BOND -- A BASE PAIR THAT DOES NOT JOIN PROPERLY. THIS ACTION CREATES A **"BUMP"** -- A MALFORMATION IN THE MOLECULAR STRUCTURE. **PROOFREADING DNA POLYMERASE** IDENTIFIES THE MISMATCHED BASE AND REPLACES IT WITH THE APPROPRIATE COMPLEMENT.

ANY MISTAKES THAT GET BY THE PROOFREADING DNA POLYMERASE ARE CAUGHT BY **YET ANOTHER** SYSTEM OF SPECIALIZED ENZYMES THAT EFFECT MISMATCH REPAIR **AFTER** REPLICATION IS COMPLETE.

WHEN THE **FRAGMENTS** HAVE BEEN **JOINED** AND ANY **MISTAKES** HAVE BEEN **CORRECTED**, THE NEW DNA WINDS INTO NUCLEOSOMES AND ASSUMES THE DEFAULT **CHROMOSOMAL** FORM.

HISTONE

HERE WE GO AGAIN...

OH -- ONE LAST DETAIL -- TO TOP OFF REPLICATION, THE **ENDS** OF THE NEW CHROMOSOMES MUST BE FINISHED. THESE ENDS -- CALLED **TELOMERES** -- ARE COMPRISED OF **SINGLE-STRANDED NUCLEOTIDE SEQUENCES** THAT DON'T CODE FOR PHYSICAL TRAITS AND ARE COMMONLY, IF UNFAIRLY, REFERRED TO AS **JUNK DNA**.

JUNK?

FEH! NATURE ISN'T THAT WASTEFUL.

JUNK DNA ACTUALLY HAS MANY USES, AND IN THIS CASE IT SIMPLY PROTECTS THE EXTREME ENDS OF THE DNA FROM DAMAGE. TELOMERES MUST BE **FINISHED** TO BE EFFECTIVE, BUT ARE FINISHED DIFFERENTLY IF THE REPLICATION PROCESS IS FOR **SIMPLE** CELL REPRODUCTION-- CALLED **MITOSIS** -- OR FOR **SEX** CELL REPRODUCTION--WHICH IS CALLED **MEIOSIS**.

PERSONALLY, I DON'T CARE FOR THE MITOSIS SCENARIO.

IN **MITOSIS**, ANOTHER ONE OF THOSE SPECIALIZED ENZYMES SIMPLY **EVENS UP** THE STRANDS OF JUNK DNA FORMING THE TELOMERES...

SNIP!

SEE WHAT I MEAN?

TELOMERASE

WE OFFER EXTRA PROTECTION!

...BUT IN **MEIOSIS**, WHICH IS RESPONSIBLE FOR PASSING A COMPLETE SET OF CHROMOSOMES DOWN TO THE ORGANISM'S OFFSPRING, YET ONE MORE PROTECTIVE MEASURE IS TAKEN AND THE SINGLE STRAND IS FINISHED AS A STABLE DOUBLE-STRAND HELIX WITH THE AID OF **RNA PRIMER** AND THE ENZYME **TELOMERASE**.

AND **THAT** IS HOW, WITH **EXCRUCIATING** ATTENTION TO FIDELITY, DNA TAKES CARE OF **ITSELF**.

BUT HOW DOES IT TAKE CARE OF THE GREATER ORGANISM? HOW DOES IT MAKE **PHYSICAL** ITS **CODED INFORMATION**?

EXACTLY WHAT I'VE BEEN WAITING TO HEAR...

WELL, I'M HERE TO TELL YOU THAT DNA DOES IT WITH THE HELP OF ITS MUCH LESS CAUTIOUS, MUCH MORE **PROMISCUOUS** COUSIN...

...YOU GUESSED IT... RNA!

How **TITILLATING!** It sounds— indiscreet.

ACTUALLY, IT'S ALL **VERY** DISCREET— AND CAREFULLY ARRANGED.

WHILE DNA STAYS SAFE AND SECURE IN THE NUCLEUS, BOLD RNA TAKES THE CHANCES, VENTURING OUT INTO THE CELL'S CYTOPLASM— THE FLUID THAT SURROUNDS THE NUCLEUS—

—TO MINGLE WITH FREE-RANGING, EAGER-TO-BOND BUILDING MATERIALS...

...AND TO MINGLE WITH A WHOLE **NEW** CREW OF **CONSTRUCTION-MINDED** ENZYMES, AND PROTEIN-MAKING FACTORIES CALLED **RIBOSOMES!** RNA CARRIES DNA'S INSTRUCTIONS TO HERE, THE CYTOPLASM, WHERE GENETIC CODE BECOMES **PHYSICAL REALITY.**

!

mRNA

EVERY GENE GETS **TRANSLATED** AND **TRANSCRIBED** INTO ITS SPECIFIC **POLYPEPTIDE,** OTHERWISE KNOWN AS A **PROTEIN—** THE BUILDING BLOCKS OF LIFE ON EARTH! EVERY **GENE** BLUEPRINTS A **SPECIFIC POLYPEPTIDE!**

SURE, IT'S A LOT OF WORK, BUT IT'S A **LIVING.**

rRNA

tRNA

CYS

ARG

TRP

LYS

GLY

THR

BUT— BEFORE WE GET INTO ALL **THAT,** WE NEED TO GO BACK INTO THE NUCLEUS AND THE BLUEPRINT ITSELF, DNA.

THAT'S RIGHT— IT ALL STARTS WITH ME.

SPECIFIC MESSENGERS TRIGGER SPECIFIC GENES AT SPECIFIC TIMES.

THIS IS KNOWN AS **GENE EXPRESSION.**

CHEMICAL MESSENGERS INCOMING— SOMEONE NEEDS TO EXPRESS!

DNA from a HUMAN PERSPECTIVE
PART 1

THOMAS HUNT MORGAN

HUMANS HAVE BEEN AWARE FOR **MILLENNIA** OF THE REALITY OF GENETICS AND HEREDITY, BUT THE **MECHANICS** BEHIND LIFE'S ABILITY TO PASS PHYSICAL CHARACTERISTICS FROM ONE GENERATION TO THE NEXT WERE UNCOVERED ONLY WITHIN THE LAST **CENTURY.**

GREGOR MENDEL'S SEMINAL CONCEPTUAL BREAKTHROUGH OF 1865 -- THAT A PROPERTY CALLED **GENES** CONVEYED HERITABLE TRAITS -- TOOK ALMOST HALF A CENTURY TO GAIN ACCEPTANCE. AND IT TOOK THAT MUCH TIME **AGAIN** TO DISCOVER JUST EXACTLY **WHAT** "GENES" WERE. THE WORK OF **MANY** HUMANS, SLOWLY PUTTING PIECE AFTER PIECE IN PLACE, WAS NEEDED TO SOLVE THAT PUZZLE.

CHROMOSOMES WERE RECOGNIZED AS UNITS OF HEREDITY IN 1903. IN 1910 THE BRILLIANT **THOMAS HUNT MORGAN** WAS ABLE TO DEMONSTRATE, BY BREEDING GENERATION AFTER GENERATION OF MUTATED FRUIT FLIES, THAT THE MYSTERIOUS GENES WERE CARRIED ON CHROMOSOMES AND WERE, INDEED, THE MECHANICAL BASIS OF HEREDITY.

OSWALD AVERY

IN 1928, BRITISH MEDICAL OFFICER **FREDERICK GRIFFITH** EXPERIMENTED WITH PNEUMONIA BACTERIA AND DISCOVERED WHAT HE CALLED THE **TRANSFORMING PRINCIPLE** -- THE ABILITY OF SOME UNKNOWN MOLECULE TO PASS PHYSICAL TRAITS FROM ONE STRAIN TO ANOTHER.

OSWALD AVERY, COLIN MacLEOD, AND **MACLYN McCARTY** CONTINUED GRIFFITH'S EXPERIMENTS AND IN 1944 FINALLY PUT A FACE TO THE ELUSIVE GENE. BY SUCCESSFULLY ELIMINATING MOLECULAR COMPONENTS FROM GRIFFITH'S BACTERIA THEY WERE ABLE TO PINPOINT **DNA** AS THE TRANSFORMING PRINCIPLE -- THE MOLECULAR AGENT OF HEREDITY.

ERWIN CHARGAFF

NOT THAT EVERYONE **ACCEPTED** DNA AT FIRST. COMMON WISDOM HAD IT THAT DNA -- UNLIKE, SAY, A PROTEIN -- WAS NOT COMPLEX ENOUGH TO CARRY ALL THE GENETIC INFORMATION NECESSARY TO AN ORGANISM.

BUT, IN 1950, **ERWIN CHARGAFF'S** RESEARCH STRONGLY SUGGESTED THAT DNA'S FOUR BASES WORKED IN COMPLEMENTARY COMBINATIONS, ALLOWING FOR MUCH GREATER COMPLEXITY THAN PREVIOUSLY CONSIDERED POSSIBLE.

AND SO DNA -- A MOLECULE THAT WAS ULTIMATELY DISCOVERED IN 1869 BY SWISS SCIENTIST FRIEDRICH MIESCHER -- WAS FINALLY ACCEPTED AS THE BOTTOM-LINE STUFF OF LIFE.

BUT DNA **ITSELF** STILL NEEDED TO BE UNDERSTOOD.

REMEMBER — ALMOST **EVERY** TYPE OF CELL OF EVERY ORGANISM CARRIES THAT ORGANISM'S **COMPLETE GENOME** — ALL ITS GENETIC INSTRUCTIONS. SO, IF THESE CELLS CARRY ALL GENES...

RED BLOOD CELLS ARE PRIVILEGED CHARACTERS — THEY DON'T BOTHER TO CARRY ALL THOSE GENETIC INSTRUCTIONS.

...WHY DOESN'T SKIN SPROUT **EYEBALLS** — WHY DON'T LIVERS GROW **TOES**? WHY DO THE APPROPRIATE CHARACTERISTICS DEVELOP ONLY IN THE APPROPRIATE ORGANS AT THE APPROPRIATE TIME?

BECAUSE GENES ARE CALLED ON TO **EXPRESS** — TO BECOME **ACTIVE** — ONLY IN CERTAIN CELLS AT CERTAIN TIMES. THEY ARE SWITCHED **ON** OR **OFF** AS THEY ARE NEEDED.

GENE ON

GENE OFF

IN MOST CELLS, THE DEFAULT SETTING IS "OFF." WHY? BECAUSE THE **UNRESTRICTED** GROWTH OF A PARTICULAR PHYSICAL CHARACTERISTIC CAN BE BAD.

NO KIDDING...

CANCER IS THE RESULT OF AN UNCONTROLLED GENE EXPRESSION.

IN THE EARLY STAGES OF AN ORGANISM'S DEVELOPMENT, MOST GENES BECOME **TISSUE SPECIFIC**. THAT MEANS THAT A SPECIFIC CELL'S DNA **PERMANENTLY TURNS OFF** ALL ITS GENES THAT DO NOT CODE FOR PHYSICAL TRAITS SPECIFICALLY NEEDED FOR **THAT** CELL.

BUT BEFORE THAT, ALL CELLS ARE **TOTIPOTENT** — WE CAN BECOME ANY TISSUE NEEDED!

BUT WHAT ARE THE FACTORS THAT ACTIVATE A CELL'S "TISSUE APPROPRIATE" GENES WHEN THEY ARE NEEDED?

WE'VE GOT INCOMING!

THERE ARE SEVERAL, SUCH AS...

...ENVIRONMENTAL FACTORS -- CONDITIONS GENERATED OUTSIDE THE ORGANISM. MOST IMPORTANT, PERHAPS, ARE THE EFFECTS **HEAT** AND **LIGHT** HAVE ON GENE EXPRESSION.

MAKES ME TURN **EXPRESSLY** IN THIS DIRECTION.

ALSO, CERTAIN GENES ARE **PRESET** TO ACTIVATE AT CERTAIN STAGES OF AN ORGANISM'S DEVELOPMENT. THEY RECEIVE THEIR CUES FROM CHANGES WITHIN THE CELL OR FROM NEIGHBORING CELLS.

PASS THE WORD -- TIME TO EXPRESS!

AND THEN THERE ARE **HORMONES** -- COMPLEX CHEMICALS PRODUCED BY THE BRAIN AND VARIOUS GLANDS THAT RUSH THROUGH THE CIRCULATORY SYSTEM, BRINGING THE IN-STRUCTIONS TO ACTIVATE AND DEACTIVATE GENES.

WE DO **MORE** THAN JUST GET ADOLESCENTS ALL HOT AND BOTHERED...

Oh, my...

WHEN A GENE **DOES** GET SWITCHED **ON**, A PROCESS CALLED **TRANSCRIPTION** BEGINS -- THAT GENE BECOMES "TRAN-SCRIBED," OR DUPLICATED, INTO A **MESSENGER RNA** MOLECULE.

A **RECORD** OF THE GENE'S BASE-PAIR CODE -- THE EXACT SEQUENCE OF BASES -- IS BUILT OUT OF RNA.

REMEMBER: RNA IS A **NUCLEIC ACID** LIKE ITS COUSIN DNA, BUT ITS SUGAR -- RIBOSE -- CARRIES ONE MORE OXYGEN ATOM THAN DNA'S DOES. THAT DOESN'T SEEM LIKE A BIG DIFFERENCE, BUT THE EXTRA O MAKES RNA MUCH FRIENDLIER, MUCH MORE PRONE TO CANOODLING WITH OTHER MOLECULES THAN DNA.

RIBOSE

DEOXYRIBOSE

AND THAT'S AN **IMPORTANT** DISTINCTION, AS WE'LL SEE.

URACIL

THYMINE

ANOTHER IMPORTANT DISTINCTION: RNA'S **BASES** ARE SLIGHTLY DIFFERENT FROM DNA'S. WHILE BOTH EMPLOY GUANINE, CYTOSINE, AND ADENINE, RNA USES **URACIL (U)** INSTEAD OF THYMINE.

URACIL IS ACTUALLY A CLOSE RELATIVE OF THYMINE AND TAKES T'S PLACE AS A COMPLEMENT TO A. BUT IT'S A MUCH MORE REACTIVE MOLECULE.

IT BONDS EASILY WITH **ALL THE OTHER BASES**...

SO I BOND EASILY -- THAT DOESN'T MEAN I'M **LOOSE**...

...GIVING A STRAND OF RNA THE ABILITY TO CONTORT INTO DIFFERENT NON-LINEAR SHAPES. THESE VARIOUS SHAPES -- CALLED **SECONDARY STRUCTURES** -- HELP RNA COMPLETE ITS MANY TASKS.

... I'M JUST **FLEXIBLE**.

tRNA

ONE LAST DISTINCTION: UNLIKE DOUBLE-STRANDED DNA, RNA IS IN ALMOST ALL ITS FORMS A **SINGLE-STRANDED MOLECULE**. THIS, ALONG WITH ALL THOSE OTHER DISTINCTIONS, MAKES RNA RELATIVELY **UNSTABLE**. UNLIKE DNA, WHICH MUST REMAIN INVIOLATE, THE ORGANISM NEEDS RNA TO CEASE FUNCTIONING AND **DETERIORATE** ONCE ITS TASK HAS BEEN COMPLETED.

WHAT?

mRNA

OKAY -- THAT'S THE **BACKGROUND** SKINNY ON RNA. NOW HERE'S HOW THE **TRANSCRIPTION PROCESS** FROM DNA TO RNA WORKS.

FIRST, THE GENE THAT IS TO BE EXPRESSED -- THE **SEQUENCE OF BASE UNITS** WITHIN THE DNA MACRO-MOLECULE THAT CONTAIN THE CODES FOR ACTUALIZING A SPECIFIC PHYSICAL TRAIT -- MUST BE LOCATED.

USING THE HUMAN GENOME AGAIN FOR OUR EXAMPLE, WE ARE TALKING ABOUT **ROUGHLY 3 BILLION BASE PAIRS**, OF WHICH 1% CAN BE TRANSCRIBED TO **MESSENGER RNA**.

THAT TRANSLATES TO ABOUT 25,000 GENES, ALL OF WHICH VARY GREATLY IN SIZE FROM ABOUT 3,000 BASE PAIRS TO AS MANY AS 2.5 MILLION!

TO FIND THE DESIRED GENE, A GROUP OF COOPERATIVE ENZYMES -- CALLED A **HOLOENZYME COMPLEX** -- SCANS THE DNA UNTIL THE CORRECT, IDENTIFYING **PROMOTER SEQUENCE** OF BASE PAIRS IS FOUND. THE **PROMOTER SEQUENCE** TELLS THE HOLOENZYME THAT TRANSCRIPTION FOR THAT PARTICULAR GENE NEEDS TO START **HERE**.

HOLOENZYME COMPLEX

WE KNOW IT'S HERE SOMEPLACE...

39

THE PROMOTER SEQUENCE -- OFTEN CODED "TATA," AS SHOWN HERE -- IS ALWAYS LOCATED ON THE DNA STRAND **NOT** USED AS THE TEMPLATE FOR TRANSCRIPTION TO **MESSENGER RNA (mRNA)**. IT TELLS THE ENZYMES TO TRANSCRIBE FROM THE **OPPOSITE** STRAND.

THE PROMOTER **ALSO** CONTROLS HOW **MANY TIMES** THE GENE IS TRANSCRIBED TO mRNA, BASED ON HOW MUCH PHYSICAL MATERIAL IS REQUIRED -- HOW MANY NEW COPIES OF THAT PARTICULAR GENE THE ORGANISM NEEDS.

NUTHIN' STARTS WITHOUT MY SAY-SO.

ONCE THE PROMOTER SEQUENCE HAS BEEN CORRECTLY IDENTIFIED, THE HOLOENZYME COMPLEX BINDS TO IT AND SIGNALS **RNA POLYMERASE** TO STAND BY FOR ACTION!

BUT BE GENTLE, ALL RIGHT?!

WHEN THE PROMOTER SEQUENCE SIGNALS **"READY!"** THE HOLOENZYME COMPLEX MAKES A TINY BREAK IN THE FIRST NUCLEOTIDES OF THE GENE, ALLOWING RNA POLYMERASE TO SEE PAST THE SUGAR-PHOSPHATES TO THE BASE PAIRS.

← RNA POLYMERASE

GOIN' IN!

ONCE RNA POLYMERASE CAN GET AT THE BASE PAIRS, IT BEGINS BREAKING APART A STRING OF BASE PAIRS ABOUT 20 PAIRS LONG -- THIS IS CALLED THE **TRANSCRIPTION BUBBLE**. NOW mRNA **SYNTHESIS** CAN BEGIN.

RNA POLYMERASE

COMIN' THROUGH!

UNLIKE THE DNA TRANSLATION PROCESS, RNA TRANSCRIPTION DOESN'T NEED A PRIMER TO GET ITS SHOW ON THE ROAD. IT JUST READS THE FIRST BASE OF THE TRANSCRIPTION BUBBLE AND STARTS PULLING IN FREE-FLOATING **NUCLEOTIDES** CARRYING THE APPROPRIATE COMPLEMENTARY BASES.

FREE RNA NUCLEOTIDES

THE NUCLEOTIDES FORM THEIR PHOSPHATE-TO-SUGAR BONDS AND THE PROCESS CONTINUES DOWN THE GENE. RNA POLYMERASE OPENS THE HELIX—AND CONTINUES THE TRANSCRIPTION BUBBLE AND SYNTHESIZING MORE mRNA.

AS THE mRNA STRAND GROWS, IT IS PUSHED AWAY FROM ITS TEMPLATE, AND THE DNA SNAPS CLOSED AS THE TRANSCRIPTION BUBBLE MOVES ON. THIS ONGOING TRANSCRIPTION PROCESS IS CALLED **ELONGATION.**

WHEN THE GENE IS **COMPLETELY** TRANSCRIBED, RNA POLYMERASE HITS A **TERMINATOR SEQUENCE OF BASES** THAT TELLS IT TO STOP AND TO **DETACH** FROM THE DNA TEMPLATE. THE DNA MOLECULE SNAPS BACK TO ITS USUAL DOUBLE-HELIX SHAPE, ITS JOB DONE!

WORK CONTINUES, HOWEVER, ON OUR NEW mRNA STRAND. LIKE GENOMES IN GENERAL, THERE ARE SEQUENCES OF BASES WITHIN TRANSCRIPTION UNITS THAT DON'T CODE FOR **PHENOTYPE**—PHYSICAL TRAITS, THAT IS. THIS "JUNK" IS REFERRED TO AS **INTRONS**, A FANCY NAME MEANING "INTERVENING SEQUENCES."

A CLEVERLY NAMED ENZYME CALLED A **SPLICEOSOME** RECOGNIZES THESE INTRONS AND EDITS THEM OUT. THEN IT **SPLICES** TOGETHER THE REMAINING **EXON** SEQUENCES (THOSE THAT **DO** CODE FOR A PHENOTYPE) INTO A SEAMLESS, WASTE-FREE **MESSAGE DELIVERY SYSTEM** OF DNA'S INSTRUCTIONS!

TO FINISH OFF OUR STREAMLINED mRNA STRAND, A **CAP OF GUANINE** IS ADDED TO ITS LEADING END, WHILE **ADENINES** ARE STUCK ON THE TAIL. THESE PROTECT THE UNSTABLE MOLECULE FROM DECAYING **TOO** QUICKLY...

...BECAUSE mRNA NOW HAS ITS **MISSION** TO COMPLETE. THAT INVOLVES CONVEYING THE INSTRUCTIONS IT HAS TRANSCRIBED FROM DNA OUT OF THE **NUCLEUS**...

I GOTTA GET **OUT** OF THIS PLACE...

...IF IT'S THE LAST THING I EVER DO...

...AND INTO THE **CYTOPLASM** AND A WORLD OF MOLECULAR RAW MATERIAL AND CHEMICAL FACTORIES, ALL WAITING TO CREATE NEW **PROTEIN** -- THE CONSTRUCTION MATERIAL NECESSARY TO ALL LIFE FUNCTIONS!

THERE ARE **LOTS** OF IMPORTANT THINGS FLOATING AROUND IN THE CYTOPLASM, BUT OUR IMMEDIATE CONCERN IS RESTRICTED TO **AMINO ACIDS** AND TWO OTHER RNA VARIANTS: **RIBOSOMAL RNA,** CLEVERLY REFERRED TO AS **rRNA,** AND **TRANSFER RNA,** WHICH IS CALLED -- YUP, YOU GUESSED IT! -- **tRNA.**

YEAH, THAT'S RIGHT -- WE'RE RELATED. WHAT ABOUT IT?

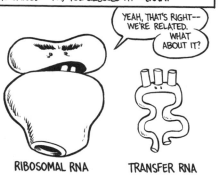

RIBOSOMAL RNA   TRANSFER RNA

AMINO ACIDS ARE THE MOLECULAR COMPONENTS THAT FORM THOSE LONG **POLYPEPTIDE** CHAINS -- **PROTEINS** AND **ENZYMES.** POLYPEPTIDES CAN BE 50 TO 1,000 AMINO ACIDS IN LENGTH, AND THAT ALLOWS FOR A GREAT VARIETY OF PROTEINS, INDEED.

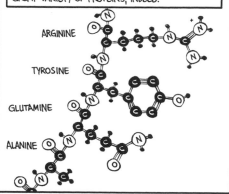

ARGININE

TYROSINE

GLUTAMINE

ALANINE

DNA'S INSTRUCTIONS FOR ASSEMBLING THE CYTOPLASM'S LOOSE AMINO ACIDS INTO A SPECIFIC PROTEIN HAVE BEEN ENCODED IN mRNA BY TRANSCRIPTION.

NOW THAT CODE MUST BE **TRANSLATED** TO THE LANGUAGE OF THE PROTEINS.

Hmmm...

RNA to PROTEIN DICTIONARY

CODON

THE SECRET OF THE CODE CARRIED BY mRNA IS THAT THE BASES MUST BE READ **THREE AT A TIME.** THIS **TRIPLET** IS CALLED A **CODON.** THE FOUR DIFFERENT BASES (A, C, G, AND U), READ IN COMBINATIONS OF ANY THREE, YIELD A TOTAL OF **64 POSSIBLE CODONS.**

THESE 64 DIFFERENT CODONS CODE FOR THE 20 DIFFERENT VARIETIES OF AMINO ACID. THERE'S A GOOD DEAL OF REDUNDANCY HERE, WITH A NUMBER OF BASE COMBINATIONS CODING FOR THE SAME PROTEIN. ACTUALLY, 61 OF THE **CODONS** CODE FOR THE **20 TYPES** OF AMINO ACIDS, WHILE THE REMAINING 3 ARE SIGNALS TO "STOP" THE TRANSLATION PROCESS.

FIRST LETTER ↓	SECOND LETTER				THIRD LETTER ↓
	U	C	A	G	
U	PHENYLALANINE PHENYLALANINE LEUCINE LEUCINE	SERINE SERINE SERINE SERINE	TYROSINE TYROSINE STOP STOP	CYSTEINE CYSTEINE STOP TRYPTOPHAN	U C A G
C	LEUCINE LEUCINE LEUCINE LEUCINE	PROLINE PROLINE PROLINE PROLINE	HISTIDINE HISTIDINE GLUTAMINE GLUTAMINE	ARGININE ARGININE ARGININE ARGININE	U C A G
A	ISOLEUCINE ISOLEUCINE ISOLEUCINE METHIONINE & START	THREONINE THREONINE THREONINE THREONINE	ASPARAGINE ASPARAGINE LYSINE LYSINE	SERINE SERINE ARGININE ARGININE	U C A G
G	VALINE VALINE VALINE VALINE	ALANINE ALANINE ALANINE ALANINE	ASPARTATE ASPARTATE GLUTAMATE GLUTAMATE	GLYCINE GLYCINE GLYCINE GLYCINE	U C A G

I KNOW WHERE YOUR AMINO ACIDS COME FROM...

IT'S INTERESTING TO NOTE THAT THESE CODONS—WHICH ARE **ALWAYS** READ IN THE DIRECTION OF mRNA'S LEADING END MOVING TOWARD ITS TAIL—CODE FOR THE SAME AMINO ACIDS IN **NEARLY EVERY** ORGANISM ON EARTH.

SO WHAT **READS** THE CODONS?

RIBOSOMES—THE rRNA FACTORIES THAT **PRODUCE** PROTEINS—DO! RIBOSOMES CAN READ AND WORK WITH ANY mRNA STRAND, AND THEY COME IN **TWO** PARTS—A **SMALL** AND A **LARGE SUBUNIT.**

THE DIVISION OF LABOR!

Time OUT!

We just ran through a **LOT** of new information. Let me see if I've got it right.

Not only can DNA **REPLICATE** itself, but when a job needs doing, DNA's instructions are **TRANSCRIBED** into another form called RNA...

PUT IT THIS WAY: WHEN **ENVIRONMENTAL** CONDITIONS OR **ORGANIZATIONAL** DECISIONS CALL FOR NEW MATERIAL FOR **MAINTAINING** OR **DEVELOPING** THE ORGANISM...

...WORKERS IN THE **DEPARTMENT** OF DNA, THE ORGANISM'S DATABANK OF BLUEPRINTS, ARE ALERTED.

THEY ASSEMBLE DNA'S BLUEPRINTS FOR THE PROPER RNA CREW TO TAKE DNA'S INSTRUCTIONS TO THE **RIBOSOME FACTORY,** WHERE THE NEEDED MATERIAL IS PRODUCED.

DNA from a HUMAN PERSPECTIVE
PART 2

DNA MAY HAVE BEEN ACCEPTED AS THE GENETIC MATERIAL BY THE EARLY 1950s, BUT ITS **STRUCTURE** -- THE SHAPE THAT ALLOWS IT TO DO ITS REMARKABLE WORK -- WAS STILL UNKNOWN.

JAMES WATSON AND **FRANCIS CRICK** BEGAN A RESEARCH PARTNERSHIP AT CAMBRIDGE UNIVERSITY IN 1951, WITH THE GOAL OF DISCOVERING THAT STRUCTURE. WATSON SPECIALIZED IN X-RAY PHOTOGRAPHY, AND HIS RESULTS WERE LEADING THE TWO TO BELIEVE THAT DNA'S OVERALL SHAPE WAS THAT OF A HELIX.

NOT FAR AWAY, AT KING'S COLLEGE, LONDON, **ROSALIND FRANKLIN** BEGAN SIMILAR RESEARCH THAT SAME YEAR WITH **MAURICE WILKINS**. HER EXCELLENT X-RAY PHOTOGRAPHY OF DNA ALSO INDICATED A POSSIBLE HELIX SHAPE, ALTHOUGH SHE WAS CAUTIOUS ABOUT PUBLICLY ANNOUNCING HER FINDINGS.

WATSON AND CRICK INITIALLY THOUGHT THE STRUCTURE OF DNA PLACED THE SUGAR/PHOSPHATE BACKBONE AT THE **CENTER** OF THE MACROMOLECULE. BUT, AFTER ATTENDING A LECTURE BY FRANKLIN FEATURING HER RESEARCH INDICATING THE OPPOSITE, THEY WERE CONVINCED TO RETHINK THEIR MODEL AND CORRECTLY REPOSITIONED THE SUGAR/PHOSPHATE COMPONENT OF DNA AS A FIRM BUT FLEXIBLE OUTER SKIN, WITH THE FOUR BASES ATTACHING ON THE INSIDE OF IT.

WILKINS ALSO SUPPOSEDLY ALLOWED WATSON TO VIEW FRANKLIN'S UNPUBLISHED RESEARCH PAPERS. THIS UNETHICAL ACTION WOULD CERTAINLY HAVE FURTHERED WATSON AND CRICK'S RACE TO DISCOVER DNA'S STRUCTURE FIRST. THE CHARGE REMAINS CONTROVERSIAL.

WATSON AND CRICK APPLIED THEIR RESEARCH-GUIDED THEORIES TO MODELS CONSTRUCTED OF CARDBOARD AND PASTE, REARRANGING MOLECULAR PLACEMENTS UNTIL THEY EVENTUALLY LOCKED TOGETHER IN LOGICAL ATOMIC BONDS. TOO MANY PIECES OF THE PUZZLE FIT FOR THEIR MODEL TO BE WRONG -- THE NOW-FAMILIAR **DOUBLE-HELIX** STRUCTURE OF DNA WAS REVEALED FOR THE FIRST TIME!

WATSON AND CRICK, ALONG WITH WILKINS, WERE AWARDED THE **NOBEL PRIZE FOR MEDICINE** IN 1962.

FRANKLIN, WHO HAD CONTRIBUTED MUCH TO THE DISCOVERY, WAS NOT ELIGIBLE FOR THE DISTINCTION. SHE DIED IN 1958 OF CANCER, A GENETICALLY RELATED DISEASE THAT SOMEDAY MAY BE BANISHED AS OUR UNDERSTANDING OF DNA GROWS.

JAMES WATSON

FRANCIS CRICK

ROSALIND FRANKLIN

THAT PRODUCTION IS CALLED THE **TRANSLATION PROCESS.** IT BEGINS WHEN A **SMALL RIBOSOME SUBUNIT** LATCHES ONTO **ANY STRAND** OF mRNA THAT HAS ENTERED THE CYTOPLASM.

SMALL RIBOSOME SUBUNIT

mRNA

THE mRNA IS CONVEYED THROUGH THE SMALL SUBUNIT UNTIL THE **START CODON**—WHICH IS ALWAYS THE TRIPLET "**AUG**"—IS LOCATED. WHEN THE SMALL SUBUNIT READS **AUG**, IT SENDS A SIGNAL CALLING THE APPROPRIATE MOLECULE OF **TRANSFER RNA.**

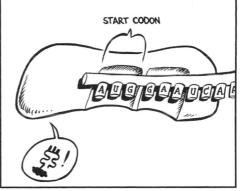

START CODON

AUGGAAUCAA

≈≋ !

tRNA MOLECULES ARE THE **DELIVERY SYSTEMS** OF THE TRANSLATION PROCESS— THE **TRUCKERS**, IF YOU WILL. THEY HAUL THE APPROPRIATE AMINO ACID TO THE RIBOSOME, AS SPECIFIED BY THE CODON BEING READ.

SER

ARG

LYS

tRNA IS VERY SPECIALIZED, HAVING CONTORTED INTO A UNIQUE SHAPE NECESSARY TO ITS FUNCTION -- THANKS TO THE BONDING ABILITIES OF **URACIL.** EACH tRNA MOLECULE CARRIES A TRIPLET OF BASES CALLED AN **ANTI-CODON,** WHICH IS **COMPLEMENTARY** TO ITS SPECIFIC CODON ON mRNA.

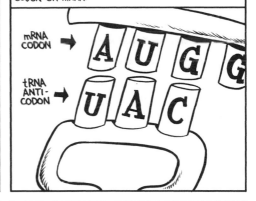

mRNA CODON → A U G G

tRNA ANTI-CODON → U A C

A SPECIALIZED ENZYME READS tRNA'S ANTI-CODON, **INFERS** ITS **CODON** FROM THAT, AND ATTACHES ITS FREIGHT—THE SPECIFIED **AMINO ACID** -- TO tRNA'S **ACCEPTOR ARM.**

U A C

AMINO ACID

SO, WHEN THE RIBOSOME READS AUG—"**START!**"—OFF THE mRNA, IT CALLS FOR A tRNA MOLECULE CARRYING THE **AMINO ACID fMET** TO LINK WITH THAT CODON.

AUGGAAUCAA

UAC

fMET

fMET IS THE ONLY AMINO ACID THAT CAN **BEGIN** A POLYPEPTIDE CHAIN. NOW THAT IT IS IN PLACE, A **LARGE RIBOSOME** SUBUNIT JOINS WITH THE **SMALL**, AND THE CREATION OF A PROTEIN BEGINS...

LARGE RIBOSOME SUBUNIT

...WITH THE **ELONGATION PROCESS**. THE **SECOND CODON** ON THE mRNA STRAND IS READ AT A **SECOND SITE** WITHIN THE MERGED RIBOSOME, WHICH CALLS FOR A **tRNA** CARRYING **THAT** SPECIFIED AMINO ACID. THIS tRNA, LIKE THE FIRST, LOCKS ITS ANTI-CODON TO THE mRNA'S CODON.

THE RIBOSOME TAKES THE AMINO ACID THAT THIS SECOND tRNA HAS CARRIED WITH IT AND **BONDS** IT TO THE FIRST AMINO ACID, fMET. AND ONCE ITS AMINO ACID IS UNLOADED, tRNA IS DISCONNECTED FROM mRNA AND RELEASED OUT AN **EXIT SITE** TO PICK UP ANOTHER AMINO ACID FOR DELIVERY.

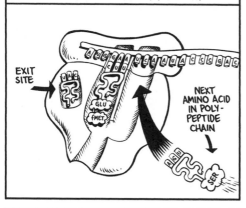

EXIT SITE

NEXT AMINO ACID IN POLY-PEPTIDE CHAIN

AND SO ELONGATION GOES, WITH THE RIBOSOME SHUFFLING THE mRNA STRAND THROUGH, READING CODONS, AND ATTACHING THE APPROPRIATE AMINO ACIDS IN A GROWING POLYPEPTIDE CHAIN-- UNTIL IT HITS A "STOP" CODON.

"STOP"

NO tRNA HAS AN ANTI-CODON TO MATCH "STOP," AND SO THE PROCESS OF ELONGATION ENDS. HELPFUL PROTEINS CALLED **RELEASE FACTORS** RECOGNIZE "STOP" AND CUT THE NEW POLYPEPTIDE CHAIN **FREE** OF THE RIBOSOME SUBUNITS.

SNIP!

THE TWO RIBOSOMAL SUBUNITS SEPARATE...

...AND A NEW PROTEIN IS BORN, READY TO **CONTORT** INTO ITS UNIQUE SHAPE, **JOIN** WITH OTHER PROTEINS, AND GO ABOUT ITS PARTICULAR BUSINESS OF **DEVELOPING** OR **MAINTAINING** THE GREATER ORGANISM!

I **THINK** I follow you...

FROM DNA TO PROTEIN:

1. First, during **TRANSCRIPTION**, a string of **mRNA** is created based on the instructions of **DNA**.

2. Next, **SPLICEOSOMES** clean up the mRNA, getting rid of the **JUNK**.

3. The mRNA leaves the **NUCLEUS** and enters the cell's **CYTOPLASM**...

4. ...where it meets up with a **RIBOSOME** to begin the **TRANSLATION** process.

5. The ribosome, with the help of **tRNA**, uses the mRNA's instructions to build a **CHAIN OF AMINO ACIDS**.

6. This chain folds into a **PROTEIN** — the building block of life!

CYTOPLASM

NUCLEUS

EUKARYOTE CELL

Please tell me that's all there is to it!

ALMOST, SIRE!

RIBOSOMES AND tRNA, MEANWHILE, ARE FREE TO LOOK FOR OTHER mRNA STRANDS AND **CONTINUE** THE PROCESS OF BUILDING MORE POLYPEPTIDES. THEY ARE BOTH VERY **RECYCLABLE**.

OKAY, WHO'S NEXT? WHO'S NEXT?!

LYS ARG

fMET

TYR

mRNA, HOWEVER, IS MADE TO DIE YOUNG. IT DELIVERS DNA'S INSTRUCTIONS TO THE DEGREE REQUIRED, AND THEN **DISINTEGRATES**, A GOOD SOLDIER TO THE END.

IT WAS FUN WHILE IT LASTED...

mRNA

AND **THAT** -- **WHEW!** -- CONCLUDES THE **MOLECULAR** PORTION OF MY REPORT.

mRNA

...SIRE?...

...YOUR ROYAL PERCEPTIVENESS?

Mind... spinning...

... At least **THREE** of my five brains must be overheated...

I KNOW, SIRE. I **KNOW.**

BUT, REST ASSURED, I'VE **FRONT-LOADED** MY REPORT WITH THE MOST TECHNICALLY CHALLENGING INFORMATION.

THE FOUNDATIONS OF THE LIFE PROCESS ARE DIFFICULT TO GRASP, INDEED...

...BUT A BASIC UNDERSTANDING OF DNA AND RNA AND WHAT THEY DO IS NECESSARY TO FULLY APPRECIATE THE **REALLY GOOD STUFF.**

I PROMISE YOU, THINGS GET A BIT LESS ARCANE FROM HERE ON, AS WE BEGIN TO LOOK AT...

...**SEX!**

CHAPTER 2

How the System Works--Sex and the Cellular Life

NOW THAT WE HAVE A BASIC UNDERSTANDING OF HOW GENETIC PROCESSES WORK ON A **MOLECULAR** LEVEL...

...WE CAN APPLY THAT KNOWLEDGE AS WE REFOCUS OUR ATTENTION TO THE **CELLULAR** LEVEL. CELLS ARE LARGELY COMPOSED OF WATER AND **PROTEINS** MANUFACTURED BY RNA IN RIBOSOME FACTORIES FROM **DNA** INSTRUCTIONS.

CELLS ARE THE MOST BASIC **LIVING** UNITS OF ALL ORGANISMS. SOME ORGANISMS CONSIST OF A SINGLE CELL, WHILE OTHERS ARE IMMENSELY COMPLICATED COMPOSITES OF MANY **COOPERATIVE** CELLS.

TO UNDERSTAND **REPRODUCTION** AND **HEREDITY** IN **ANY** ORGANISM, WE MUST FIRST UNDERSTAND THAT THESE PROCESSES OCCUR ON A CELLULAR BASIS.

CELLS, AND THUS ALL EARTHLY LIFE, ARE BROKEN INTO TWO **DOMAINS.** *PROKARYOTES* ARE THAT GROUP OF ORGANISMS THAT **LACK** A **CELLULAR NUCLEUS** AND HAVE NEVER DEVELOPED BEYOND THE SINGLE-CELL STAGE...

BACTERIA

...WHILE *EUKARYOTES* ARE THAT GROUP OF ORGANISMS WHOSE CELLS CONTAIN A MEMBRANE-BOUND NUCLEUS.

PROTISTS AND ALL MULTI-CELLULAR ORGANISMS

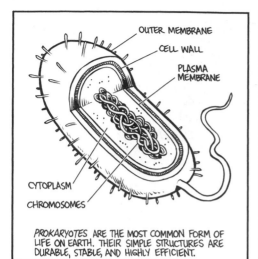

OUTER MEMBRANE
CELL WALL
PLASMA MEMBRANE
CYTOPLASM
CHROMOSOMES

PROKARYOTES ARE THE MOST COMMON FORM OF LIFE ON EARTH. THEIR SIMPLE STRUCTURES ARE DURABLE, STABLE, AND HIGHLY EFFICIENT.

EUKARYOTE ANATOMY GETS A GOOD DEAL MORE COMPLEX.

PROKARYOTE **DNA** IS RELATIVELY SIMPLE, AS WELL. THE CHROMOSOMES GENERALLY TAKE **LOOPED** SHAPES, FLOATING FREELY IN THE CELL'S CYTOPLASM...

...WHILE EUKARYOTE DNA IS MUCH MORE COMPLICATED. ITS CHROMOSOMAL DEFAULT FORM IS THAT OF LONG, STRINGY **SEGMENTS**, WHICH ARE PROTECTED FROM POTENTIAL DAMAGE BY BEING WITHIN THE NUCLEUS.

PROKARYOTES ARE **ASEXUAL**-- THEY REPRODUCE THROUGH BINARY FISSION, THE LITERAL SPLITTING OF THE PARENT CELL INTO TWO IDENTICAL NEW ONES.

BACTERIA HAVE NEVER BEEN CONSIDERED SEXY...

SMOOCH!

EUKARYOTES CAN BE ASEXUAL AS WELL, BUT THEY HAVE ALSO DEVELOPED THE INNOVATION OF **SEXUAL REPRODUCTION**, WHERE THE DNA OF **TWO GENDERS** --THE **MALE** AND THE **FEMALE** -- MUST BE **COMBINED** TO CREATE OFFSPRING.

EUKARYOTE CELL

RIBOSOMES

CENTRIOLES

MITOCHONDRION

GOLGI APPARATUS

MICRO-TUBULES

LYSOSOME

PLASMA MEMBRANE

FLAGELLUM

NUCLEUS

CHROMOSOMES

ENDOPLASMIC RETICULUM

AS WE CAN SEE HERE, THE INNARDS OF A TYPICAL EUKARYOTE CELL ARE QUITE COMPLEX, WITH MANY DIFFERENT PARTS -- CALLED **ORGANELLES** -- WORKING TOGETHER TO BUILD PROTEIN, TO SYNTHESIZE ENERGY...

...AND TO REPRODUCE.

OF PARTICULAR INTEREST ARE THE **MITOCHONDRIA** -- ORGANELLES THAT CONVERT FOOD TO ENERGY AND POWER THE CELL LIKE LITTLE BATTERIES. THEY ALSO CONTAIN THEIR VERY OWN DNA, WHICH CODES AND DEVELOPS NEW MITOCHONDRIA **INDEPENDENT** OF THE GREATER CELL'S DNA.

MITOCHONDRION

PLANT CELLS CONTAIN **ADDITIONAL** ORGANELLES CALLED **CHLOROPLASTS** THAT CONDUCT THE **PHOTOSYNTHESIS** PROCESS UNIQUE TO GREEN PLANTS. CHLOROPLASTS, TOO, CARRY THEIR OWN DNA.

CHLOROPLAST

THE UNIQUE DNA IN MITOCHONDRIA AND CHLOROPLASTS SUGGESTS THAT THESE ORGANELLES WERE ONCE INDEPENDENT **PROKARYOTES** THAT FORMED **SYMBIOTIC** RELATIONSHIPS WITH A GREATER PROTIST BODY IN THE DISTANT PAST.

EUKARYOTE CELLS COMPRISING MULTI-CELLULAR ORGANISMS ARE DIVIDED INTO **TWO TYPES**: THE **SOMATIC** CELLS, WHICH FORM THE **BODY** OF THE ORGANISM, PERFORMING THE GRUNT WORK OF CONSTANT BUILDING AND MAINTAINING...

BUSY, BUSY, BUSY...

...AND THE **SEX** CELLS, WHOSE SOLE PURPOSE IS TO **REPRODUCE** THE ORGANISM -- TO CREATE A NEW GENERATION.

SOME CELLS HAVE **ALL** THE FUN.

BECAUSE SOMATIC CELLS MUST PERFORM ALL THE MANY DIVERSE FUNCTIONS NECESSARY TO GROW, MAINTAIN, AND REPAIR AN ORGANISM, THEY CARRY THE ORGANISM'S **COMPLETE** GENOME WITH THEM, EXPRESSING THE GENES REQUIRED FOR THEIR SPECIALIZATION.

FAT

LEAF

BONE

THEY REPRODUCE THEMSELVES BY SIMPLE **MITOSIS**— THE PROCESS THAT DUPLICATES **EXACTLY** THE CELL'S COMPLETE SET OF CHROMOSOMES AND PASSES IT UNALTERED FROM ONE GENERATION TO THE NEXT.

BONE

SAME BONE

SEX CELLS, ON THE OTHER HAND, HAVE A SPECIAL **PROBLEM** TO OVERCOME, AS THE SEXUAL REPRODUCTION STRATEGY REQUIRES THAT THE GENETIC MATERIAL OF **TWO** INDIVIDUAL CELLS BE COMBINED.

PROBLEM? WHAT PROBLEM?

SPECIFICALLY, SEX CELLS MUST FIND A WAY TO **HALVE** THEIR FULL COMPLEMENT OF DNA, SO THAT WHEN COMBINED, A CORRECT, PRECISE COMPLEMENT OF DNA IS FORMED FROM THE TWO CELLS.

THIS PROCESS OF HALVING THE SEX CELL'S DNA IS CALLED **MEIOSIS**.

ALL IT TAKES IS **MILLIONS** OF YEARS OF EXPERIENCE.

BEFORE WE GO ANY FURTHER, WE NEED TO TAKE A CLOSER LOOK AT **CHROMOSOMES** — THOSE PHYSICALLY ORGANIZED FORMS OF DNA THAT CARRY THE CELL'S GENES.

EVERY EUKARYOTE CARRIES ITS GENETIC HERITAGE IN A NUMBER OF CHROMOSOMES THAT IS **STANDARD** TO ITS SPECIES.

MAIZE CARRIES 20 CHROMOSOMES...

...AND A **DOG**, 78.

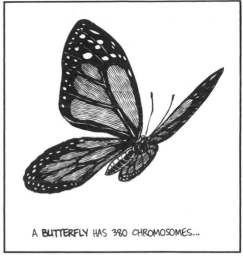

A **BUTTERFLY** HAS 380 CHROMOSOMES...

...WHILE THAT INTELLIGENT BIPED, THE **HUMAN**, HAS A COMPLEMENT OF 46.

THE NUMBER OF CHROMOSOMES AN ORGANISM BEARS HAS NOTHING TO DO WITH ITS SIZE OR BIOLOGICAL COMPLEXITY.

17 18

22 X/Y

REGARDLESS OF THE NUMBER, ALL EUKARYOTE CHROMOSOMES BREAK DOWN INTO **TWO TYPES**. **GONOSOMES** -- SEX CHROMOSOMES -- CARRY THE GENES THAT DETERMINE AN ORGANISM'S **REPRODUCTIVE CHARACTERISTICS**.

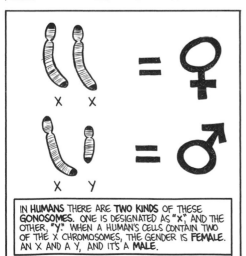

X X = ♀

X Y = ♂

IN **HUMANS** THERE ARE **TWO KINDS** OF THESE **GONOSOMES**. ONE IS DESIGNATED AS "X," AND THE OTHER, "Y." WHEN A HUMAN'S CELLS CONTAIN TWO OF THE X CHROMOSOMES, THE GENDER IS **FEMALE**. AN X AND A Y, AND IT'S A **MALE**.

THE REMAINING 44 CHROMOSOMES **NOT** DEVOTED TO SEXUAL REPRODUCTION ARE CALLED **AUTOSOMES**.

THE 44 AUTOSOMES ARE ACTUALLY **TWIN SETS OF 22 CHROMOSOMES.** IN OTHER WORDS, HUMANS HAVE 22 UNIQUE NON-SEX CHROMOSOMES, EACH OF WHICH HAS A **DUPLICATE.**

EACH OF THE CHROMOSOMES IN A PAIRED SET OF DUPLICATES IS CALLED A **HOMOLOG,** AND THE TWO CHROMOSOMES' RELATIONSHIP IS DESCRIBED AS **HOMOLOGOUS.**

MY HOMEY!

THIS BIOLOGICAL STRATEGY OF CARRYING TWO COPIES OF CHROMOSOMES IS A COMMON ONE AMONG EARTHLY *EUKARYOTES*--AND THOSE EMPLOYING IT ARE TERMED **DIPLOID.**

BUT IT IS NOT THE ONLY GAME IN TOWN...

THE DIPLOID CLUB

...THERE ARE FERNS, FOR INSTANCE, THAT CARRY AS MANY AS 630 PAIRS OF CHROMOSOMES! SOME BEES, ANTS, AND WASPS, ON THE OTHER EXTREME, ARE **HAPLOID** ORGANISMS-- MEANING THEY CARRY ONLY **ONE** SET OF CHROMOSOMES.

AND WE DO JUST FINE, THANK YOU.

WHICH BRINGS US BACK AGAIN TO OUR **HUMAN** EXAMPLE-- WHILE HUMAN AUTOSOME CHROMOSOMES ARE DIPLOID, THE **GONOSOMES** -- THE TWO SEX CHROMOSOMES -- ARE HAPLOID.

I AM A UNIQUE INDIVIDUAL.

AS AM I. ONE OF A KIND.

THERE IS A VERY IMPORTANT REASON FOR **HAPLOIDY** IN SEX CELLS, AS WE'LL COME TO SEE, BUT FIRST...

...WE NEED TO LOOK AT THE PHYSICAL COMPONENTS OF CHROMOSOMES.

CHROMOSOMES ARE COMPOSED OF **DNA** AND THE **PROTEINS** THAT HELP PACKAGE THE DNA INTO MANAGEABLE SHAPES.

MOST IMPORTANT, CHROMOSOMES CARRY **GENES**, THOSE ORGANIZED SEQUENCES OF DNA THAT **DO** SUPPLY THE INSTRUCTIONS FOR CREATING PHYSICAL TRAITS.

IN THEIR USUAL, **DEFAULT** STATE, EUKARYOTE CHROMOSOMES WITHIN THE NUCLEUS ARE VERY LONG, STRINGY SEGMENTS OF LINEAR DNA THAT ARE INVISIBLE EVEN THROUGH A GOOD MICROSCOPE...

...BUT DURING MITOSIS OR MEIOSIS, CHROMOSOMES MORPH INTO A DISTINCTLY **ROBUST, PLUMPER** SHAPE THAT IS MUCH EASIER TO SEE THROUGH A MICROSCOPE, AND MUCH EASIER TO ILLUSTRATE.

SO, AS A VISUAL AID, WE WILL STICK WITH THE MITOSIS/MEIOSIS FORM AS WE LOOK AT **CHROMOSOME ANATOMY**...

THIS IS A SINGLE CHROMOSOME. EACH GENE IS FOUND AT A SPECIFIC LOCATION, CALLED A **LOCUS**, ALONG THE LENGTH OF THE CHROMOSOME. THE LOCUS OF A GENE **NEVER** VARIES WITHIN THE CHROMOSOME.

THE PINCHED AREA BETWEEN THE TWO ARMS IS CALLED THE **CENTROMERE**, AND VARIES IN EXACT LOCATION AMONG DIFFERENT CHROMOSOMES.

TELOMERES CAP THE TWO ENDS WITH **NONCODING DNA** THAT SERVES AS A BUFFER TO PROTECT THE GENES.

WHEN A CHROMOSOME REPLICATES ITSELF, USING THE DNA REPLICATION PROCESS WE LOOKED AT EARLIER, THE NEWLY FORMED CHROMOSOME REMAINS CONNECTED TO THE ORIGINAL AT THE CENTROMERE. THE TWO ARE NOW CALLED **SISTER CHROMATIDS**.

SISTER CHROMATIDS

NOW WE CAN GET INTO AN ASPECT OF CHROMOSOME ANATOMY THAT IS OF **CENTRAL** IMPORTANCE TO SEXUAL REPRODUCTION!

THESE TWO SISTER CHROMATIDS ARE **HOMOLOGOUS.** THEY ARE **DIPLOID AUTOSOMES,** MEANING THEY BEAR IDENTICAL COPIES OF ALL CHROMOSOMES **NOT** RELATED TO SEX. THE GENES THEY CARRY SIT AT THE SAME LOCI ON EITHER HOMOLOG.

BUT-- THOSE SAME GENES ARE NOT NECESSARILY **IDENTICAL!** EACH HOMOLOG MAY HAVE **VARIANTS** OF THE SAME GENE. THESE ARE CALLED **ALLELES,** AND ARE INDICATED IN NOTATION BY THE USE OF THE UPPER- OR LOWERCASE OF THE SAME LETTER.

— HOMOLOGS —

A GENE THAT CODES FOR A PARTICULAR PHYSICAL TRAIT-- LET'S SAY HAIR COLOR-- CAN COME IN **TWO OR MORE** VARIANTS.

ONE ALLELE OF THAT HAIR-COLOR GENE MAY CODE FOR **BROWN**...

BB

...WHILE THE OTHER ALLELE PRODUCES **BLOND** HAIR.

IT IS THE **ALLELES** THAT ALLOW FOR THE GREAT **VARIATION** WE SEE WITHIN ORGANISMS THAT REPRODUCE SEXUALLY!

bb

NOW, UNDERSTAND THAT THIS EXAMPLE IS EXTREMELY **SIMPLIFIED**...

...MOST OF AN INDIVIDUAL ORGANISM'S PHYSICAL TRAITS ARE CREATED THROUGH THE INTERACTION OF A **NUMBER** OF ALLELES, AS WELL AS ENVIRONMENTAL FACTORS...

bb

...But I'm starting to get the picture.

All these chromosome complexities create variety within a particular species of organism.

WHEN APPLIED TO SEXUAL RE-PRODUCTION, YES -- THEY DO!

IN FACT, WE'VE NOW GOT ENOUGH BACKGROUND INFORMATION TO FINALLY START LOOKING AT **REPRODUCTIVE TECHNIQUE**...

The **GOOD** stuff!

...BEGINNING WITH **MITOSIS**...

...THE PROCESS BY WHICH **SOMATIC CELLS** DUPLICATE THEMSELVES...

...PASSING ON IDENTICAL SETS OF THEIR CHROMOSOMES TO A NEW GENERATION OF CELLS.

MITOSIS IS PART OF THE **CELL CYCLE** -- THE PHASES OF LIFE A CELL GOES THROUGH, ONE DIVISION TO THE NEXT...

IT'S OUR DAILY ROUTINE, EXCEPT WE DON'T GET TO SLEEP...

...AND ALTERNATES WITH A PART OF THE LIFE CYCLE CALLED **INTERPHASE**, WHICH IS THE TIME FOR A CELL TO **FEED, GROW,** AND **REPLICATE** ITS DNA.

--WE'RE GREAT MULTI-TASKERS.

THE INTERPHASE BREAKS DOWN INTO **THREE STAGES**. FIRST COMES A PERIOD IN WHICH A NEW CELL **GROWS** RAPIDLY. THE VORACIOUS LITTLE THING EATS AND BREATHES, GAINING STRENGTH AND ENERGY.

MMMMM... MY **FAVORITE** PERIOD.

←PROTEIN

AT THIS POINT SOME TYPES OF SOMATIC CELLS PULL **OUT** OF THE CYCLE. RED BLOOD CELLS AND, GENERALLY SPEAKING, MUSCLE CELLS ARE AMONG THOSE THAT NEVER GO FURTHER -- THEY NEVER REPLICATE.

WE RED BLOOD CELLS ARE **FAR** TOO BUSY TO FIND THE TIME TO REPRODUCE.

MOST CELLS, HOWEVER, **CONTINUE** THROUGH THE CYCLE TO THE **SECOND STAGE**, IN WHICH THE **REPLICATION** OF THE CELL'S **NUCLEAR DNA** TAKES PLACE. ALL THE CHROMOSOMES ARE DUPLICATED, BECOMING **SISTER CHROMATIDS** CONNECTED AT THEIR CENTROMERES.

AFTER DNA REPLICATION IS COMPLETE, THE CELL TAKES ANOTHER SHORT BREATHER, GROWING SOME MORE—THE **THIRD** STAGE OF METAPHASE—BEFORE FINALLY ENTERING...

WHEW!

IT TAKES A LOT OF ENERGY TO BE A GOOD CELL...

...MITOSIS...

...THE **SEPARATION** OF THE 46 CONJOINED SISTER CHROMATIDS INTO **TWO CELLS**.

THERE ARE FIVE **STAGES** TO THIS PROCESS.

AND, BY THE WAY, **PLANT** AND **ANIMAL** MITOSIS **DIFFER** A BIT, SO WE'RE GOING TO STICK WITH THE **ANIMAL** PROCESS HERE.

FIRST COMES **PROPHASE**, IN WHICH THE CHROMOSOMES CONTRACT AND CONDENSE INTO THEIR ROBUST, PLUMP, VISIBLE SHAPES.

PROPHASE

TO 19TH-CENTURY SCIENTISTS, THE FIRST OBSERVERS OF CHROMOSOMES, THIS CHANGE IN FORM SEEMED TO **POP** THESE MICROSCOPIC SQUIGGLES INTO EXISTENCE FROM OUT OF NOWHERE!

THERE WAS NOTHING THERE A MINUTE AGO!

AT THE SAME TIME THAT THE CHROMOSOMES COMPACT INTO THEIR NEW SHAPES, THE CELL NUCLEUS BEGINS TO **DISINTEGRATE**, ALLOWING THE CHROMOSOMES TO MIGRATE INTO OPEN **CYTOPLASM**.

FREE AT LAST!

NEXT COMES **METAPHASE**, IN WHICH THE WANDERING CHROMOSOMES LINE UP LIKE GOOD SOLDIERS ALONG THE **EQUATOR**, AN IMAGINARY CENTERLINE THROUGH THE MIDDLE OF THE CELL...

METAPHASE

... WHILE AT THE CELL'S **POLES**, OPPOSITE THE EQUATOR, TWO ORGANELLES CALLED **CENTRIOLES** PROJECT THREAD-LIKE **PROTEIN SPINDLES**...

... WHICH **GRAB** EACH LINED-UP CHROMOSOME AT ITS **CENTROMERE**. THE SPINDLES BEGIN TO PULL THE SISTER CHROMATIDS APART...

...AND THE PROCESS ENTERS **ANAPHASE** AS THE SPINDLES DRAW A FULL SET OF CHROMOSOMES TO EACH OF THE POLES.

ANAPHASE

THEN COMES **TELOPHASE**, IN WHICH A NEW **NUCLEAR MEMBRANE** BEGINS TO FORM PROTECTIVELY AROUND EACH COMPLETE SET OF CHROMOSOMES AS THEY UNCOIL BACK INTO THEIR DEFAULT FORM.

TELOPHASE

FINALLY, MITOSIS CONCLUDES WITH **CYTOKINESIS**: THE ORIGINAL CELL FINISHES DIVIDING AROUND THE TWO NEW NUCLEI AND THE TWO DAUGHTER CELLS ARE NOW READY TO BEGIN THE CYCLE AGAIN.

CYTOKINESIS

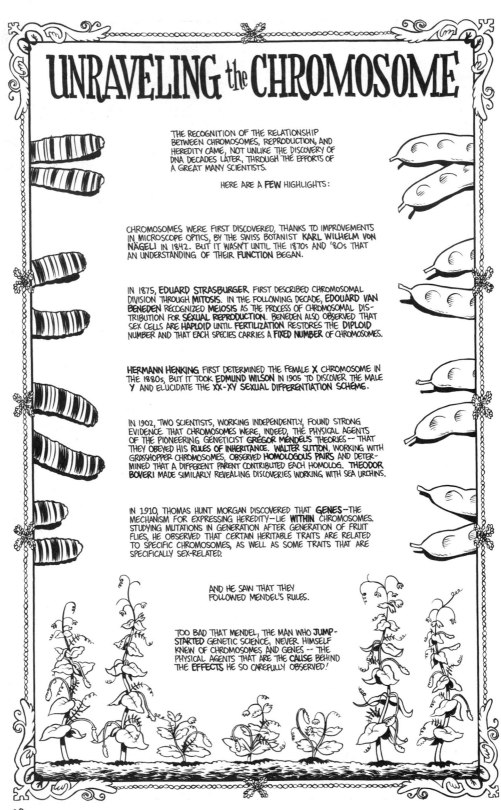

UNRAVELING the CHROMOSOME

THE RECOGNITION OF THE RELATIONSHIP BETWEEN CHROMOSOMES, REPRODUCTION, AND HEREDITY CAME, NOT UNLIKE THE DISCOVERY OF DNA DECADES LATER, THROUGH THE EFFORTS OF A GREAT MANY SCIENTISTS.

HERE ARE A **FEW** HIGHLIGHTS:

CHROMOSOMES WERE FIRST DISCOVERED, THANKS TO IMPROVEMENTS IN MICROSCOPE OPTICS, BY THE SWISS BOTANIST **KARL WILHELM VON NÄGELI** IN 1842. BUT IT WASN'T UNTIL THE 1870s AND '80s THAT AN UNDERSTANDING OF THEIR **FUNCTION** BEGAN.

IN 1875, **EDUARD STRASBURGER** FIRST DESCRIBED CHROMOSOMAL DIVISION THROUGH **MITOSIS**. IN THE FOLLOWING DECADE, **EDOUARD VAN BENEDEN** RECOGNIZED **MEIOSIS** AS THE PROCESS OF CHROMOSOMAL DISTRIBUTION FOR **SEXUAL REPRODUCTION**. BENEDEN ALSO OBSERVED THAT SEX CELLS ARE **HAPLOID** UNTIL **FERTILIZATION** RESTORES THE **DIPLOID** NUMBER AND THAT EACH SPECIES CARRIES A **FIXED NUMBER** OF CHROMOSOMES.

HERMANN HENKING FIRST DETERMINED THE FEMALE **X** CHROMOSOME IN THE 1880s, BUT IT TOOK **EDMUND WILSON** IN 1905 TO DISCOVER THE MALE **Y** AND ELUCIDATE THE **XX-XY** SEXUAL DIFFERENTIATION **SCHEME**.

IN 1902, TWO SCIENTISTS, WORKING INDEPENDENTLY, FOUND STRONG EVIDENCE THAT CHROMOSOMES WERE, INDEED, THE PHYSICAL AGENTS OF THE PIONEERING GENETICIST **GREGOR MENDEL'S** THEORIES -- THAT THEY OBEYED HIS **RULES OF INHERITANCE**. **WALTER SUTTON**, WORKING WITH GRASSHOPPER CHROMOSOMES, OBSERVED **HOMOLOGOUS PAIRS** AND DETERMINED THAT A DIFFERENT PARENT CONTRIBUTED EACH HOMOLOG. **THEODOR BOVERI** MADE SIMILARLY REVEALING DISCOVERIES WORKING WITH SEA URCHINS.

IN 1910, THOMAS HUNT MORGAN DISCOVERED THAT **GENES**—THE MECHANISM FOR EXPRESSING HEREDITY—LIE **WITHIN** CHROMOSOMES. STUDYING MUTATIONS IN GENERATION AFTER GENERATION OF FRUIT FLIES, HE OBSERVED THAT CERTAIN HERITABLE TRAITS ARE RELATED TO SPECIFIC CHROMOSOMES, AS WELL AS SOME TRAITS THAT ARE SPECIFICALLY SEX-RELATED.

AND HE SAW THAT THEY FOLLOWED MENDEL'S RULES.

TOO BAD THAT MENDEL, THE MAN WHO **JUMP-STARTED** GENETIC SCIENCE, NEVER HIMSELF KNEW OF CHROMOSOMES AND GENES -- THE PHYSICAL AGENTS THAT ARE THE **CAUSE** BEHIND THE **EFFECTS** HE SO CAREFULLY OBSERVED!

The cell cycle-- MITOSIS.

I like it. It's stable and predictable-- like the tides.

Like US.

YES, MITOSIS IS THE BEST GUARANTEE THAT EVERY NEW GENERATION WILL REMAIN THE SAME.

VERY FEW VARIABLES CAN ENTER INTO THE PROCESS. THERE IS LITTLE OPPORTUNITY FOR CHANGE--

--FOR MUTATION...

...AND OUR SIMILARLY STABLE REPRODUCTIVE STRATEGY HAS FOR AGES WORKED VERY WELL FOR SQUINCHDOM.

BUT THAT SAME STABILITY CAN BE A LIABILITY WHEN UNSTABLE ENVIRONMENTAL FACTORS CALL FOR A RAPID, ADAPTIVE RESPONSE.

Such as the... GENETIC DIFFICULTIES we're now facing...

YES.

THE SOLUTION MIGHT JUST BE A REPRODUCTIVE PROCESS THAT GIVES SEXUAL ORGANISMS AN ADVANTAGE WE LACK...

...MEIOSIS! IT'S THE SPECIALIZED PROCESS THAT DERIVES FROM MITOSIS BUT, UNLIKE MITOSIS, HALVES THE DNA IN SEX CELLS TO 23 CHROMOSOMES. THIS ALLOWS TWO PARENT SEX CELLS...

HI! I'M MOM...

...AND I'M DAD!

...TO EACH CONTRIBUTE 23 CHROMOSOMES AND FORM A NEW, COMPLETELY UNIQUE OFFSPRING WITH A FULL COMPLEMENT OF 46 CHROMOSOMES!

WE'RE GOING TO CALL HIM A ZYGOTE.

IT'S A FAMILY NAME.

REMEMBER-- THE 44 **NON-SEX-RELATED** HUMAN CHROMOSOMES ARE **DIPLOID**. MEIOSIS DIVIDES EACH INTO TWO **HAPLOID** SETS...

IT'S LIKE I'M BEING TORN APART FROM **MYSELF!**

...WHILE SEPARATING THE TWO HAPLOID **GONOSOMES**-- THE SEX CHROMOSOMES-- AS WELL.

SO WHAT? I NEVER FELT **WE** HAD MUCH IN COMMON.

SO--

WHILE **MITOSIS** PRODUCES **TWO CELLS** WITH **IDENTICAL** DNA CONTENT...

...MEIOSIS PRODUCES **FOUR** CELLS, EACH WITH **HALF** THE FULL COMPLEMENT OF DNA.

MEIOSIS **FURTHER** ENHANCES GENETIC VARIABILITY BY RE-COMBINING ALLELES--

-- IT **EXCHANGES** GENE VARIANTS ON **HOMOLOGOUS** CHROMOSOMES!

THIS I've got to see.

IN ADDITION TO ALL THAT, THE TWO INDIVIDUALS CONTRIBUTING THEIR SEX CELLS TO THE REPRODUCTIVE PROCESS MUST BE OF A DIFFERENT **GENDER.**

HI! I'M MOM...

...AND I'M DAD!

ONE MUST BE A **FEMALE**, WHOSE SEX CELLS ARE CALLED **EGGS**, AND THE OTHER A **MALE**, WHOSE SEX CELLS ARE CALLED **SPERM.**

Two genders--

Both contributing genetic material that's been halved and mixed around...

It's all so very-- **ALIEN!**

BUT IT'S NOT SO HAPHAZARD AS IT MAY SEEM TO YOU NOW. LET ME EXPLAIN...

...THE **TWO** ROUNDS OF MEIOSIS!

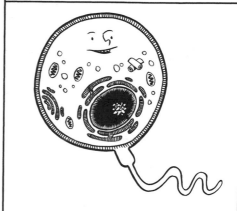

JUST LIKE MITOSIS, **MEIOSIS I** IS PRECEDED BY A PERIOD OF **INTERPHASE** DURING WHICH THE SEX CELL GROWS, REPLICATES ITS DNA, AND PREPARES FOR DIVISION.

MEIOSIS I **BEGINS** WITH **PROPHASE I**, IN WHICH THE **HOMOLOGOUS CHROMOSOMES** — EACH SET BEING THE CONTRIBUTION OF A DIFFERENT PARENT-- **PAIR UP.**

PROPHASE I

THEN THE AMAZING PROCESS OF **RECOMBINATION** OCCURS: THE HOMOLOGOUS CHROMOSOMES ACTUALLY JOIN TOGETHER AND SPECIALIZED ENZYMES BEGIN TO CUT OUT **BLOCKS** OF ALLELES...

... AND **SWITCH** THEM BETWEEN THE TWO HOMOLOGS! THIS EXCHANGE OF GENE VARIANTS BETWEEN CHROMOSOMES IS CALLED **CROSSING-OVER.**

SWITCH!

THE TRADED ALLELES ARE ALL PLACED IN THEIR CORRECT LOCI -- EVERY GENE STAYS IN ITS PROPER ORDER -- BUT ARBITRARILY SELECTED ALLELES HAVE BEEN SWITCHED BETWEEN CHROMOSOMES!

eyelash length
shoe size
earlobe type
eye color
blood type
hair color
HEMOPHILIA
HAIRLINE
NOSE SHAPE
FINGER LENGTH

EYELASH LENGTH
SHOE SIZE
EARLOBE TYPE
EYE COLOR
BLOOD TYPE
HAIR COLOR
hemophilia
hairline
nose shape
finder length

THE RECOMBINED HOMOLOGS, **STILL CONNECTED,** NOW ENTER **METAPHASE I.** THEY LINE UP AS PAIRS ACROSS THE CELL EQUATOR...

METAPHASE I

...AND, IN **ANAPHASE I,** ARE PULLED APART BY SPINDLES TO THE CENTRIOLES AT EITHER OPPOSITE POLE. ONE COMPLETE SET OF CHROMOSOMES GOES TO EACH POLE.

ANAPHASE I

IT'S PURE LUCK OF THE DRAW AS TO WHICH HOMOLOG IS PULLED TO WHICH POLE.

EACH PAIRED CHROMOSOME IS SEPARATED **INDEPENDENTLY** OF THE OTHERS. THIS **INDEPENDENT ASSORTMENT** IS YET ANOTHER LAYER OF RANDOMNESS THAT CONTRIBUTES TO EVERY SEX CELL'S UNIQUE MIX OF ALLELES.

IN **TELOPHASE I,** THE CELL GOES THROUGH ITS **FIRST ROUND OF DIVISION,** PRODUCING **TWO CELLS,** EACH WITH A FULL COMPLEMENT OF CHROMOSOMES.

TELOPHASE I

THEN, AFTER A SHORT PERIOD OF REST, **MEIOSIS II** BEGINS...

WHEW-- AT **LAST** I CAN RELAX MY CHROMOSOMES...

HUH! SOMATIC CELLS HAVE IT EASY.

...AS THE TWO CELLS PREPARE FOR A **SECOND DIVISION.**

ALREADY?!

SIGH-- TIME TO TIGHTEN UP AGAIN...

IN **METAPHASE II**, THE CHROMOSOMES OF BOTH CELLS AGAIN OBEDIENTLY LINE UP ALONG THEIR EQUATORS, AND THE SPINDLES **AGAIN** GRAB THEM AROUND THEIR CENTROMERES.

METAPHASE II

IN **ANAPHASE II**, THE SPINDLES PULL THE CHROMOSOMES APART. THEY SEPARATE THE **SISTER CHROMATIDS**, DRAWING A SET OF EACH TO OPPOSITE POLES.

ANAPHASE II

FINALLY, IN **TELOPHASE II**, NUCLEAR MEMBRANES FORM AROUND THE **FOUR SETS OF CHROMOSOMES**.

TELOPHASE II

CELL DIVISION IS COMPLETED AND MEIOSIS HAS PRODUCED **FOUR SEX CELLS**, EACH **HAPLOID** WITH A ONE-HALF COMPLEMENT OF 23 CHROMOSOMES.

EACH CELL HAS ONE SEX CHROMOSOME. EACH HAS A RANDOM, UNIQUE RECOMBINATION OF ALLELES.

AND THAT, IN A NUTSHELL, IS **MEIOSIS**, THE GREAT TOOL OF **GENETIC DIVERSITY**.

Okay-- I think I understand the **PROCEDURE** as it takes place within **ONE INDIVIDUAL'S** CELLS...

But this **GENDER** business...

...Males and females--

--how do the two genders get their sex cells **TOGETHER**?

WELL, THE PHYSICAL INTERACTION THAT INTRODUCES MALE AND FEMALE DNA TAKES ON MANY FORMS AND BEHAVIORAL VARIATIONS. SOME ARE QUITE--AHEM--SPECTACULAR.

BUT THAT'S A FIELD OF STUDY OUTSIDE THE SCOPE OF THIS REPORT--

SO I'LL KEEP MY EXPLANATION STRICTLY CELLULAR...

GAMETOGENESIS IS THE PROCESS THROUGH WHICH MALE AND FEMALE GAMETES--SEX CELLS--ARE DEVELOPED AND MADE READY FOR FERTILIZATION--THE CREATION OF A NEW GENERATION!

HI! I'M MOM...

...AND I'M DAD!

MEIOSIS IS A PART OF GAMETOGENESIS. THE SEX CELLS THAT MEIOSIS PRODUCES IN MALES ARE CALLED SPERM, AND THOSE PRODUCED IN FEMALES ARE CALLED EGGS.

FOR GAMETOGENESIS, MALES PRODUCE SPECIAL CELLS, SPERMATOGONIA, IN THAT PART OF THEIR SEX ORGAN CALLED THE TESTES. THE SPERMATOGONIA EACH HAVE A FULL DIPLOID SET OF 46 CHROMOSOMES.

REMEMBER ME?

THROUGH MEIOSIS, THE SPERMATOGONIA ARE DIVIDED DOWN TO FOUR SPERM CELLS, EACH A HAPLOID WITH THE 23 CHROMOSOME HALF-SHARE.

NOW I'M... US!

AND BECAUSE THE COMBINATION OF SEX CHROMOSOMES THAT MAKES A HUMAN **MALE** ARE AN X AND A Y, MEIOSIS IN MALE GAMETES GUARANTEES THAT **ONE-HALF** THE SPERM PRODUCED WILL CARRY AN X AND **ONE-HALF** WILL CARRY A Y.

THE **FEMALE** PRODUCES HER DIPLOID **OOGONIA** CELL IN HER OVARIES. LIKE THE MALE EQUIVALENT, IT IS RENDERED DOWN, BY MEIOSIS, TO **FOUR HAPLOID EGGS**. BUT...

...**UNLIKE** IN SPERM PRODUCTION, ONLY **ONE** OF THE FOUR EGGS MATURES INTO A VIABLE REPRODUCTIVE CELL.

THAT'S OKAY, HONEY...

THIS IS BECAUSE OF THE HUGE AMOUNTS OF **CYTOPLASM** NEEDED TO NOURISH THE CELLULAR BEGINNINGS OF A NEW GENERATION. THE FEMALE STRATEGY IS TO **POOL** ITS CYTOPLASM INTO ONE WELL-STOCKED EGG, LEAVING THE OTHER THREE DORMANT.

...YOU GO ON WITHOUT US...

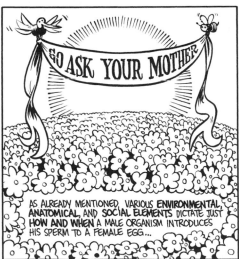

GO ASK YOUR MOTHER

AS ALREADY MENTIONED, VARIOUS **ENVIRONMENTAL, ANATOMICAL,** AND **SOCIAL ELEMENTS** DICTATE JUST **HOW AND WHEN** A MALE ORGANISM INTRODUCES HIS SPERM TO A FEMALE EGG...

...BUT ONCE THEY **ARE** INTRODUCED, WHEN THE CONDITIONS ARE **FAVORABLE**, ONE OF THE MANY DEPLOYED SPERM ENTERS THE SINGLE EGG AND **FERTILIZES** IT...

LATER FOR YOU, LOSERS!

...WHICH MEANS THAT THE TWO HAPLOID SEX CELLS -- **THE EGG AND THE SPERM** -- MERGE TO BECOME ONE NEW DIPLOID CELL CALLED A **ZYGOTE**.

THIS IS THE GENESIS OF A COMPLETELY UNIQUE INDIVIDUAL MADE FROM THE COMBINED CHROMOSOMES OF ITS MOTHER AND FATHER!

So that's why meiosis needs to **HALVE** the chromosomes in each sex cell...

...so that the mother and father can **EQUALLY CONTRIBUTE** to creating a full set in their offspring!

RIGHT! AND IF THE MALE BRINGS AN **X SEX CHROMOSOME** TO THE ZYGOTE, ITS COMBINATION WITH THE **FEMALE'S X** EQUALS A NEW **FEMALE**...

But if the sperm brings a **Y**...

...its combination with the female would produce a **MALE!**

I THINK HE'S **GOT** IT!

So, I can see how Earthly creatures utilizing sexual reproduction have opted for **VARIATION** and **DIVERSITY** as being more favorable to their survival than **STABILITY** and **RESISTANCE TO MUTATION**.

But it **STILL** seems to me that so much could go **WRONG** -- every reproduction such a **GAMBLE**...

YES, THINGS **CAN** GO WRONG -- IT IS A **GAMBLE**, OF SORTS.

But it wouldn't have lasted if it wasn't a **WINNING** strategy.

YES. AND BEYOND THAT...

...**HUMANS** HAVE FOUND WAYS OF PLAYING THE PERCENTAGES AS THEY'VE LEARNED TO READ THE LAWS OF INHERITANCE.

68

CHAPTER 3

How the System Works -- Everyone Gets an Inheritance

SO -- UP UNTIL NOW WE'VE BEEN DELVING INTO THE INCREDIBLY DIVERSE GENETIC POSSIBILITIES THAT COME WITH **SEXUAL REPRODUCTION**...

...THROUGH A LOOK AT THE **MOLECULAR** AND **CELLULAR** PROCESSES THAT MAKE THAT SYSTEM WORK.

I KNOW, I KNOW...

...NOT AS EXCITING AS THE PREVIOUS PANEL...

NOW WE'LL LOOK AT **TRANSMISSION GENETICS** --

-- THE **PATTERNS** AND **LAWS** BY WHICH PHYSICAL CHARAC- TERISTICS ARE PASSED FROM ONE GENERATION TO THE NEXT...

LAWS?

I **SWEAR** WE'RE OVER 21 AND LEGALLY MARRIED!

...OTHERWISE KNOWN AS **HEREDITY**.

OH. RIGHT.

NEVER MIND.

AS WE'VE SEEN, THOSE WONDERFUL TRAIT VARIATIONS FOUND IN SEXUALLY REPRODUCING ORGANISMS ARE POSSIBLE BECAUSE A GENE CAN COME IN DIFFERENT **VARIATIONS**...

...CALLED **ALLELES**.

SWITCH!

WE'LL SEE HERE JUST **HOW** WHAT WE'VE LEARNED ABOUT THE **CROSSING-OVER** OF ALLELES IN THE MEIOSIS PROCESS IS OF **PRACTICAL** EFFECT ON SUCCESSIVE GENERATIONS.

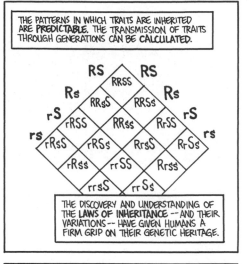

THE PATTERNS IN WHICH TRAITS ARE INHERITED ARE **PREDICTABLE**. THE TRANSMISSION OF TRAITS THROUGH GENERATIONS CAN BE **CALCULATED**.

THE DISCOVERY AND UNDERSTANDING OF THE **LAWS OF INHERITANCE** -- AND THEIR VARIATIONS -- HAVE GIVEN HUMANS A FIRM GRIP ON THEIR GENETIC HERITAGE.

EVEN AS AN OVERVIEW, IT'S HARD TO DISCUSS THIS PRACTICAL, APPLICABLE UNDERSTANDING OF THE RESULTS OF GENETICS WITHOUT INTRODUCING THE INSIGHTFUL HUMAN WHO FIRST RECOGNIZED THOSE LAWS...

...GREGOR MENDEL.

THE CONCEPT OF HERITABLE TRAITS PASSED FROM PARENT TO CHILD HAD, OF COURSE, BEEN OBSERVED AND UNDERSTOOD AT **LEAST** SINCE HUMANS FIRST BEGAN LIVING IN FAMILY GROUPS.

HE LOOKS JUST LIKE YOU!

...AND, NO DOUBT, THE ISSUE OF PATERNITY HAS **ALWAYS** BEEN OF GREAT IMPORTANCE FOR ANY SEXUAL ORGANISM RECOGNIZING THAT **LIKE BEGETS LIKE**...

HE LOOKS **NOTHING** LIKE ME!

...BUT, UNTIL MENDEL, ALL THEORIES OF HEREDITY WERE ESSENTIALLY JUST SHOTS IN THE DARK, UNSUPPORTED BY SOLID EVIDENCE. MENDEL WAS THE FIRST TO MAKE AN ACCURATE, SCIENTIFIC ANALYSIS OF HEREDITY BY MEANS OF PAINSTAKING, CONTROLLED OBSERVATION.

SO WHEN WILL THE **PATERNITY TEST** BE INVENTED?

BEFORE MENDEL

IT ALL STARTED WITH "LIKE BEGETS LIKE." SOMEWHERE IN THEIR EARLY HISTORY, HUMANS RECOGNIZED THAT ORGANISMS COULD BE COUNTED ON TO PRODUCE MORE OF THE **SAME** ORGANISM -- AND **NOT** PRODUCE ANY OTHER KIND.

EARLY MAN WAS **PRACTICAL**. THE DESIRE TO DOMESTICATE ANIMALS AND PLANTS WENT HAND IN HAND WITH THE KNOWLEDGE THAT **SEX** AND **POLLINATION** PRODUCED **BIRTHS** AND FERTILE FIELDS. **SELECTIVE CROSSING** AND **BREEDING** BECAME TOOLS FOR STRENGTHENING VALUED TRAITS AND REMOVING UNBENEFICIAL ONES.

BUT THE "**FACTORS**" THAT ALLOWED FOR THE TRANS-MISSION OF **LIKE TRAITS** FROM PARENT TO OFFSPRING COULD ONLY BE **GUESSED** AT, AND THE GUESSES WERE OFTEN CAUGHT UP IN SUPERSTITION AND CONTEMPORARY SOCIAL VALUES.

FOR CENTURIES, THE NOTION THAT **BLOOD** WAS THE TRANSMITTER OF HERITABLE TRAITS WAS WIDELY ACCEPTED AS FACT. THE CHILDREN OF BIRACIAL COUPLES OFTEN CARRIED FEATURES REMINISCENT OF BOTH PARENTS' ETHNICITY, AND THIS WAS SEEN AS EVIDENCE THAT THE TWO BLOODLINES HAD **MIXED**. THE ROYALTY OF EUROPE SUPPOSEDLY CARRIED A **SUPERIOR BLUISH BLOOD** IN THEIR VEINS AND BECAME REFERRED TO AS **BLUE BLOODS**. (ALL THE INBREEDING THAT ATTEMPTED TO KEEP THAT BLOOD BLUE DIDN'T DO THEM ANY FAVORS, THOUGH.)

PHILIP IV

THE ANCIENT **GREEKS** THOUGHT THAT PER-HAPS THE REPRODUCTIVE MATERIAL WAS SECRETED FROM **ALL PARTS** OF THE ORGANISM EQUALLY, EACH PART RESPONSIBLE FOR PASSING ITS TRAITS TO THE OFFSPRING. THIS BECAME KNOWN AS **PANGENESIS** AND NO LESS A SCIENTIST THAN CHARLES DARWIN SUB-SCRIBED TO A VERSION OF THE THEORY, PROPOSING THAT "**GEMMULES**" ARE COLLECTED FROM THROUGHOUT THE BODY DURING SEX, FOR TRANSMISSION TO THE NEXT GENERATION.

DARWIN

BUT NO THEORY WAS SUPPORTED BY HARD EVIDENCE UNTIL **GREGOR MENDEL** CONDUCTED HIS CAREFULLY CONTROLLED EXPERIMENTS IN 1865. HIS EXPERIMENTS ON GARDEN PEAS LED HIM TO ACCURATELY INTERPRET THE **NATURE** AND **ACTION** OF **GENES** -- THE NOW-ACCEPTED "**FACTORS**" OF INHERITANCE-- LONG BEFORE THEY WERE ACTUALLY SEEN OR GIVEN A NAME.

MENDEL

MENDEL WAS A 19th-CENTURY AUSTRIAN MONK WITH A PARTICULAR INTEREST IN THE PLANT VARIATIONS HE SAW IN HIS GARDENS. HE WAS ENCOURAGED TO STUDY THESE VARIATIONS BY COLLEAGUES IN BOTH RELIGION AND SCIENCE.

MENDEL MADE SOME VERY GOOD CHOICES WHILE DEVELOPING HIS SCIENTIFIC STUDY. HE CHOSE TO EXPERIMENT WITH THE COMMON **PEA PLANT**, WHICH PROVED TO BE AN EXCELLENT SUBJECT FOR UNCOVERING THE ACTIONS OF **SIMPLE INHERITANCE.**

THEN MENDEL MADE AN INTELLECTUAL BREAKTHROUGH THAT SEPARATED HIS STUDY FROM PREVIOUS ATTEMPTS TO UNDERSTAND HEREDITY: HE RECOGNIZED THAT PHYSICAL TRAITS **MUST BE OBSERVED AND QUANTIFIED INDIVIDUALLY** -- NOT AS AN **AGGREGATE** OF TRAITS OR AS A **GENERATION** OF TRAITS.

HARD TO BELIEVE IT TOOK A HUMBLE MONK TO FIGURE THIS OUT.

HE MADE FORTUITOUS CHOICES IN THE **SPECIFIC TRAITS** HE DECIDED TO STUDY IN HIS PEA PLANTS, TOO. THEY WERE ALL THE RESULTS OF GENES EXPRESSING IN ONLY **TWO ALLELIC VARIATIONS.**

THE **SHAPE** OF THIS PEA IS **ROUND**...

...AND THIS ONE IS **WRINKLED**. THAT'S ALL -- ROUND OR WRINKLED.

WE'LL SEE **LATER** THAT SOMETIMES GENES CAN COME IN **MORE** THAN TWO ALLELES. BUT, BY EITHER LUCK OR CAREFUL OBSERVATION, MENDEL LIMITED HIS EXPERIMENTS TO GENES THAT EXPRESSED IN ONLY **TWO** POSSIBLE PHENOTYPES...

RABBITS ARE NOT FOR ME -- TOO CONFUSING!

...AND THIS KEPT HIS GROUNDBREAKING OBSERVATIONS RELATIVELY **SIMPLE** TO INTERPRET, WITH NONE OF THE PUZZLING COMPLEXITY THAT WOULD HAVE OCCURRED IF HE HAD CHOSEN TRAITS WITH MORE ALLELIC VARIATION.

ANY NEW SCIENCE STARTS WITH BABY STEPS...

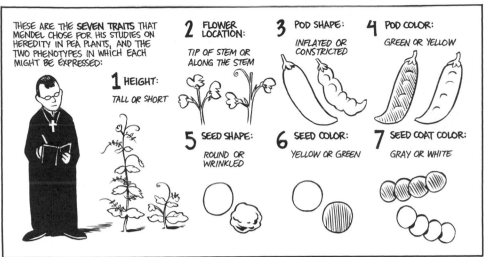

THESE ARE THE **SEVEN TRAITS** THAT MENDEL CHOSE FOR HIS STUDIES ON HEREDITY IN PEA PLANTS, AND THE TWO PHENOTYPES IN WHICH EACH MIGHT BE EXPRESSED:

1 HEIGHT:
TALL OR SHORT

2 FLOWER LOCATION:
TIP OF STEM OR ALONG THE STEM

3 POD SHAPE:
INFLATED OR CONSTRICTED

4 POD COLOR:
GREEN OR YELLOW

5 SEED SHAPE:
ROUND OR WRINKLED

6 SEED COLOR:
YELLOW OR GREEN

7 SEED COAT COLOR:
GRAY OR WHITE

MENDEL **INTERBRED** HIS PEA PLANTS -- THOUSANDS OF THEM -- GENERATION AFTER GENERATION, AND BEGAN TO **RECOGNIZE, CONFIRM,** AND **SUCCESS-FULLY PREDICT** CERTAIN **PATTERNS** OF TRAIT TRANSMISSION.

HE CAREFULLY CONTROLLED HIS EXPERIMENTS. PEA PLANTS ARE CAPABLE OF **SELF-FERTILIZATION,** AND TO ELIMINATE THAT POS-SIBILITY HE TRIMMED HIS SUB-JECT'S **ANTHERS** -- THE MALE SEX ORGAN. HE THEN CARE-FULLY INTRODUCED **POLLEN** -- THE MALE GAMETE -- TO HIS PARENT PLANTS BY HAND.

BEYOND THE ADVANTAGE OF HAVING TO DEAL WITH ONLY TWO ALLELIC VARIATIONS PER TRAIT, THE GARDEN PEA WAS ALSO A GOOD CHOICE AS IT IS A **TRUE-BREEDER.**

WHEN A TRUE-BREEDER IS ALLOWED TO **SELF-FERTILIZE,** ITS OFFSPRING INVARIABLY **RETAIN ALL THE TRAITS** DISPLAYED BY THE PARENT-- NO SURPRISES.

POLLEN

"LIKE BEGETS LIKE" IN THE **PUREST** SENSE.

AND ONE FINAL LUCKY BREAK FOR MENDEL, SOMETHING HE COULD NOT HAVE KNOWN TO CONSIDER: ALL SEVEN OF HIS CHOSEN TRAITS ARE **AUTOSOMAL** -- MEANING THEY ARE LOCATED ON **NON-SEX CHROMOSOMES.**

SEX CHROMOSOME TRAITS, WHICH MAY EXPRESS **DIFFERENTLY** DEPENDING ON GENDER, COULD ONLY HAVE MUDDIED MENDEL'S NASCENT GENETIC STUDIES.

HEY, I'M A **MONK,** FOR HEAVEN'S SAKE!

MENDEL SET UP HIS PEA PLANT EXPERIMENTS SO HE COULD OBSERVE ALL SEVEN TRAITS **INDIVIDUALLY.** TO DO THIS, HE CROSSED TRUE-BREEDERS THAT DIFFERED BY ONLY **ONE TRAIT** -- FOR INSTANCE, HE WOULD CROSS A TALL PEA PLANT WITH A SHORT, WITH THE SIX OTHER TRAITS CONSISTENT BETWEEN PARENTS.

NO DIFFERENCE OTHER THAN THE HEIGHT...

HIS RESULTS OF THESE **MONOHYBRID CROSSES** WERE EXTREMELY CONSISTENT. IN THIS CASE, THE TALL CROSSED WITH THE SHORT ALWAYS PRODUCED **TALL** OFFSPRING PLANTS.

HMM...

THAT WAS AN IMPORTANT OBSERVATION: UP UNTIL THEN, THE ACCEPTED THEORY HELD THAT INHERITED TRAITS WERE A **BLEND** OF BOTH PARENTS' TRAITS. THERE WAS CLEARLY NO BLENDING OF HEIGHT TRAITS IN THE PEA PLANTS.

WHAT MENDEL **DID** SEE WAS THAT IN THE OFFSPRING -- CALLED AN F_1 **GENERATION** -- THE TALL PARENT PEA PLANT'S TALLNESS CONSISTENTLY EXPRESSED WHILE THE SHORT PEA PLANT'S **SHORTNESS** NEVER EXPRESSED.

...WHAT HAPPENED TO THE SHORT TRAIT?!

MENDEL DIDN'T STOP HIS EXPERIMENTS HERE. HE LET THE F_1 GENERATION **SELF-FERTILIZE,** AND OBSERVED SOMETHING UNEXPECTED.

F_1 GENERATION

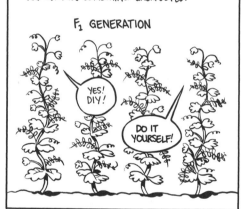

YES! DIY!

DO IT YOURSELF!

WHILE **75%** OF THE OFFSPRING OF THE F_1 GENERATION SHOWED THE TALL TRAIT...

...**25%** CAME UP **SHORT!**

THE SHORT TRAIT HAD APPARENTLY **SKIPPED** THE F_1 GENERATION, REAPPEARING IN F_2.

AND WHEN MENDEL ALLOWED THE F_2 GENERATION TO SELF-FERTILIZE...

HMM, HMM, HMM...

...HE SAW THAT THE **SHORT PEA PLANTS** OF THAT GENERATION WERE **TRUE-BREEDERS,** PRODUCING ONLY MORE SHORTS, WHILE THE OFFSPRING OF THE **TALLS** BROKE DOWN IN A **3:1 RATIO** -- THREE TALL OFFSPRING FOR EVERY SHORT.

FROM HIS OBSERVATIONS, MENDEL CONCLUDED THAT THE **UNKNOWN FACTORS** PASSING TRAITS FROM PARENT TO OFFSPRING CAME IN SETS OF **TWO VARIANTS** FOR EACH TRAIT.

IT'S CONSISTENTLY ONE OR THE OTHER!

FURTHERMORE, MENDEL SAW THAT **ONE OF THE TWO** VARIANTS OF ANY GIVEN TRAIT TENDED TO **DOMINATE** THE OTHER, **RECESSIVE**, VARIANT. THE **DOMINANT** WOULD **CONCEAL** THE PRESENCE OF THE **RECESSIVE**, ALTHOUGH THE RECESSIVE VARIANT COULD **REAPPEAR** IN SUCCESSIVE GENERATIONS.

1 THE FACTOR FOR **TALL** ALWAYS DOMINATED THE FACTOR FOR **SHORT**.

2 FLOWERS ALONG THE STEMS ALWAYS DOMINATED FLOWERS AT THE **TIPS**.

3 INFLATED PODS ALWAYS DOMINATED **CONSTRICTED** PODS.

4 GREEN PODS ALWAYS DOMINATED YELLOW PODS.

5 ROUND SEEDS ALWAYS DOMINATED **WRINKLED** SEEDS.

6 YELLOW SEEDS ALWAYS DOMINATED GREEN SEEDS.

7 GRAY SEED COATS ALWAYS DOMINATED **WHITE** SEED COATS.

THE BIOLOGICAL MECHANICS UNDERPINNING THIS WOULD, OF COURSE, NOT BE UNDERSTOOD UNTIL MUCH LATER, BUT MENDEL'S **FACTOR VARIANTS** WERE, INDEED, ALLELES -- **GENES**!

HE ACCURATELY DEDUCED THAT THE TWO PARENTS **EACH CONTRIBUTE ONE ALLELE** TO THE NEXT GENERATION.

WISH I COULD COME UP WITH A GOOD NAME FOR THESE **FACTORS**...

IF BOTH ALLELES INHERITED BY THE OFFSPRING WERE THE DOMINANT **TALL** ALLELE (T AND T), THE OFFSPRING WOULD BE TALL.

IF ONE ALLELE INHERITED WAS FOR **TALL** (T) AND THE OTHER WAS FOR **SHORT** (t), THE TALL T WOULD **DOMINATE** -- CONCEAL -- THE SHORT t AND WOULD EXPRESS AS **TALL**.

ONLY IF **BOTH** ALLELES INHERITED WERE FOR **SHORT** (t AND t) WOULD THE OFFSPRING EXPRESS AS **SHORT**.

THE TWO ALLELES INHERITED -- ONE EACH FROM BOTH PARENTS -- REPRESENT THE TRAIT'S **GENOTYPE**. IN THIS CASE, THERE ARE **THREE** GENOTYPES POSSIBLE: TT, Tt (OR tT), AND tt, AS ILLUSTRATED HERE IN A **CROSSING-SQUARE** FORMAT.

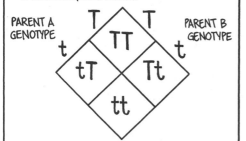

PARENT A GENOTYPE

PARENT B GENOTYPE

BUT THERE ARE ONLY **TWO PHENOTYPES** -- EXPRESSED TRAITS -- POSSIBLE. TT AND Tt BOTH EXPRESS AS TALL, WHILE tt EXPRESSES AS SHORT.

HOW ANY TWO ALLELES HAPPEN TO WIND UP TOGETHER IN THE SAME GAMETE IS ALL A MATTER OF WHAT MENDEL TERMED **SEGREGATION** AND **INDEPENDENT ASSORTMENT**. THESE PROCESSES ARE ACTUALLY FUNCTIONS OF CELLULAR MEIOSIS.

NOT THAT I KNEW ANYTHING ABOUT MEIOSIS...

REMEMBER HOW, DURING MEIOSIS, HOMOLOGOUS CHROMOSOMES **SEPARATE INDEPENDENTLY**?

A **DIPLOID** SEX CELL, WHICH HAS PAIRS OF EACH CHROMOSOME, DIVIDES INTO **HAPLOID** GAMETES -- SEX CELLS WITH ONLY ONE COPY OF EACH.

Sure I do.

And, as I recall, before they separate, the homologous chromosomes **CROSS-OVER** -- exchange -- various alleles.

EXACTLY! THEN THE MALE HAPLOID GAMETE FERTILIZES THE FEMALE HAPLOID GAMETE AND A **NEW** DIPLOID OFFSPRING, WITH A **UNIQUE COMBINATION** OF ITS PARENTS' ALLELES, IS FORMED!

THESE ARE THE **CELLULAR MECHANICS** BEHIND THE RESULTS THAT MENDEL OBSERVED --

-- A DISTRIBUTION OF ALLELES THAT ASSURES A **MAXIMUM** OF POSSIBLE VARIATION.

OUT OF ALL THOSE POSSIBLE VARIATIONS, AT LEAST **SOME** WILL BE BENEFICIAL -- INSURING THE ORGANISM'S SURVIVAL UNTIL BREEDING AGE -- AND SO SPREAD THROUGHOUT THE SPECIES IN GENERAL. THIS IS THE WAY **SEXUAL ORGANISMS** IMPROVE THEMSELVES. UNFORTUNATELY, IT'S A STRATEGY WE **ASEXUAL SQUINCH** DON'T HAVE!

NOW, TO ILLUSTRATE MENDEL'S CONCEPT OF **SEG-REGATION**, LET'S LOOK AGAIN AT HIS PEA PLANT MONOHYBRID CROSSES. THIS TIME, LET'S SAY WE CROSS A **TRUE-BREEDER** WITH **INFLATED PEA PODS**--BOTH ALLELES **DOMINANT (PP)**...

I'M HOMOZYGOUS DOMINANT--BOTH MY ALLELES ARE IDENTICAL DOMINANTS.

PP

... AND A TRUE-BREEDING PEA PLANT WITH CONSTRICTED PODS--BOTH ALLELES **RECESSIVE** (pp).

AND I'M HOMOZYGOUS RECESSIVE.

pp

GAMETES ARE HAPLOID CELLS, SO ANY GAMETE PRODUCED BY THE PP PARENT WOULD HAVE TO CARRY A **P** ALLELE.

AND ANY GAMETE PRODUCED BY THE pp PARENT WOULD, OF COURSE, BE BEQUEATHED WITH A **p**.

GAMETES

SO, THE RESULT OF CROSSING A PLANT WITH P GAMETES WITH A PLANT CARRYING p GAMETES WOULD INVARIABLY PRODUCE AN **F₁ GENERATION** OFFSPRING WITH Pp **GENOTYPES**.

I'M HETEROZYGOUS-- MY TWO ALLELES DIFFER FROM EACH OTHER.

WHEN THAT F₁ GENERATION IS ALLOWED TO SELF-FERTILIZE, EACH F₁ PEA PLANT PRODUCES AN EQUAL AMOUNT OF EITHER P OR p GAMETES, AND THE **SEGREGATION** PROCESS INHERENT IN MEIOSIS MAKES FOUR COMBINATIONS OF F₂ GENERATION OFFSPRING POSSIBLE: PP, Pp, pP, OR pp.

THIS IS A **PREDICTABLE GENOTYPIC RATIO** OF 1:2:1. 1 HOMOZYGOUS DOMINANT (PP) TO 2 HETEROZYGOUS DOMINANT (Pp OR pP) TO 1 HOMOZYGOUS RECESSIVE (pp).

AND THE **PHENOTYPIC RATIO?**

THREE INFLATED PEA POD PLANTS TO EVERY ONE CONSTRICTED PLANT.

THE **F₂ GENERATION** WILL PRODUCE **GAMETES** THAT BREAK DOWN LIKE SO:

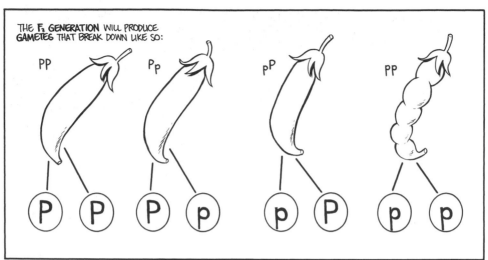

AND, IF THE F₂ GAMETES ARE ALLOWED TO SELF-FERTILIZE, WE SEE **THESE** DISTRIBUTIONS OF ALLELES INTO THE **F₃** GENERATION:

HOMOZYGOUS DOMINANT OR RECESSIVE CROSSES ARE ALWAYS TRUE-BREEDERS BUT HETEROZYGOUS CROSSES PRODUCE OFFSPRING WITH **AN EQUAL CHANCE OF HAVING ANY OF THE SAME FOUR GENOTYPE COMBINATIONS** SEEN IN THE **F₂** GENERATION.

THAT'S HOW **SEGREGATION** WORKS AMONG GENES CARRYING TWO ALLELES. IT FOLLOWS PREDICTABLE RATIOS.

Right. But I'm **SUSPICIOUS.**

This heredity business seems almost **TOO** easy. Too mathematically cut-and-dried...

YOU ARE MOST WISE AND PERCEPTIVE, INDEED. THIS IS **SIMPLE INHERITANCE** WE ARE LEARNING NOW. THINGS WILL GET MORE COMPLEX. LATER.

BUT BEFORE WE GET TO **THAT,** THERE IS ONE MORE VERY IMPORTANT MENDELIAN CONCEPT WE NEED TO EXPLORE, ONE **CLOSELY** RELATED TO SEGREGATION.

INDEPENDENT ASSORTMENT ALLOWS OFFSPRING THE OPPORTUNITY TO INHERIT ANY POSSIBLE ALLELE FROM EITHER PARENT WITHOUT CONNECTION TO, OR INFLUENCE BY, ANY OTHER ALLELE.

EVERY ALLELE IS INHERITED INDEPENDENTLY.

TO INVESTIGATE THIS PRINCIPLE, WE NEED TO MOVE, LIKE MENDEL, BEYOND MONOHYBRIDS AND INTO THE REALM OF DIHYBRIDS--

--INDIVIDUALS THAT DIFFERENTIATE BY MORE THAN ONE TRAIT.

Please-- no more PEAS!

I've never even SEEN a REAL pea and already I'm tired of them!

OKAY-- FOR VARIETY AND SIMPLICITY'S SAKE, LET'S MAKE UP OUR OWN TEST SUBJECT.

LET'S MAKE IT AN ANIMAL--A GARDEN INSECT-- THE VORACIOUS PEA BEETLE...

I SHOULD LOOK THIS HANDSOME IN REAL LIFE!

SO, TO BEGIN OUR DEMONSTRATION OF HOW TRAITS ARE SORTED IN- DEPENDENTLY OF ONE ANOTHER...

...WE'LL SAY OUR DIHYBRID SUBJECTS HAVE TWO ALLELIC VARIANCES--ONE IN CARAPACE SHAPE AND THE OTHER IN CARAPACE SPOT SIZE: A ROUND CARA- PACE (R) IS DOMINANT OVER AN OBLONG CARAPACE (r) AND BIG SPOTS (S) ARE DOMINANT OVER LITTLE SPOTS (s).

CARAPACE...?

YOUR SHELL, SLOWPOKE-- YOUR SHELL!

BOTH OF OUR PEA BEETLES ARE HOMOZYGOUS FOR THEIR TRAITS. ONE HAS AN RRss GENO- TYPE-- ROUND CARAPACE, LITTLE SPOTS. THE OTHER HAS AN rrSS GENOTYPE-- OBLONG CARAPACE, BIG SPOTS.

WE'RE SO DIFFERENT, YET I AM SO ATTRACTED TO YOU...

RRss

rrSS

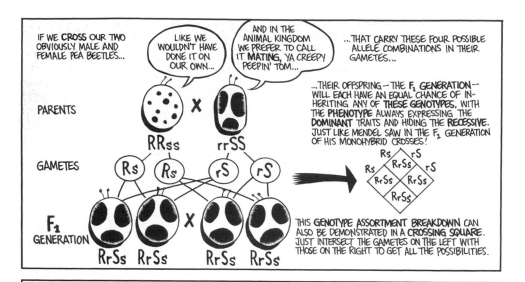

IF WE **CROSS** OUR TWO OBVIOUSLY MALE AND FEMALE PEA BEETLES...

LIKE WE WOULDN'T HAVE DONE IT ON OUR OWN...

AND IN THE ANIMAL KINGDOM WE PREFER TO CALL IT **MATING**, YA CREEPY PEEPIN' TOM...

...THAT CARRY THESE FOUR POSSIBLE ALLELE COMBINATIONS IN THEIR GAMETES...

...THEIR OFFSPRING--THE **F₁ GENERATION**-- WILL EACH HAVE AN EQUAL CHANCE OF IN-HERITING ANY OF **THESE GENOTYPES**, WITH THE **PHENOTYPE** ALWAYS EXPRESSING THE **DOMINANT** TRAITS AND HIDING THE **RECESSIVE**. JUST LIKE MENDEL SAW IN THE **F₁** GENERATION OF HIS MONOHYBRID CROSSES!

PARENTS

RRss X rrSS

GAMETES

Rs Rs rS rS

F₁ GENERATION

RrSs RrSs X RrSs RrSs

THIS **GENOTYPE** ASSORTMENT BREAKDOWN CAN ALSO BE DEMONSTRATED IN A **CROSSING SQUARE**. JUST INTERSECT THE GAMETES ON THE LEFT WITH THOSE ON THE RIGHT TO GET ALL THE POSSIBILITIES.

THE CROSSING SQUARE COMES IN HANDY IF WE LET THE F₂'S PRODUCE AN **F₂ GENERATION** WITH ALL ITS POSSIBLE **GENOTYPIC COMBINATIONS**, CLEARLY SHOWING THAT EVERY ALLELE GETS A CHANCE TO COMBINE IN EVERY COMBINATION POSSIBLE.

PHENOTYPICALLY, WHEREAS MENDEL FOUND A **3:1 RATIO** BETWEEN DOMINANT AND RECESSIVE TRAITS IN THE F₂ GENERATION OF HIS MONOHYBRID CROSSES, IN DIHYBRID CROSSES THE RATIO BECOMES **9:3:3:1**.

THAT'S THE PHENOTYPE BREAKDOWN WE SEE HERE, **NINE** PEA BEETLES WITH **BOTH** DOMINANT TRAITS (ROUND CARAPACES, BIG SPOTS) EX-PRESSED, **THREE** WITH **RECESSIVE** (OBLONG CARAPACE) AND **DOMINANT** (BIG SPOTS), **THREE** WITH **DOMINANT** (ROUND CARAPACE) AND **RECESSIVE** (SMALL SPOTS), AND **ONE** INDIVIDUAL WITH **BOTH** TRAITS RECESSIVE (OBLONG CARAPACE, SMALL SPOTS).

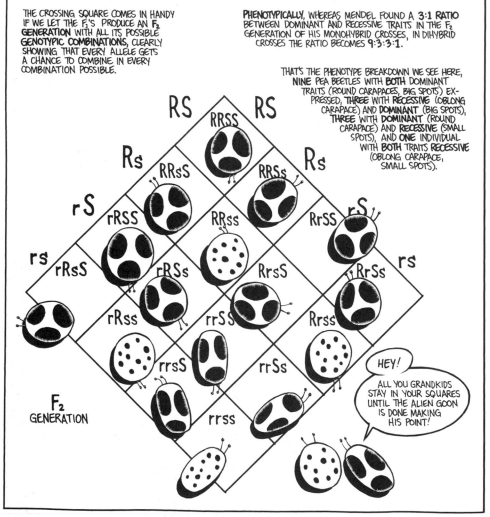

F₂ GENERATION

HEY! ALL YOU GRANDKIDS STAY IN YOUR SQUARES UNTIL THE ALIEN GOON IS DONE MAKING HIS POINT!

HIS DIHYBRID EXPERIMENTS LED MENDEL TO ACCURATELY CONCLUDE THAT **ALL TRAITS SORT OUT INDEPENDENTLY.**

ALTHOUGH ONE TRAIT MAY DOMINATE THE **EXPRESSION** OF ANOTHER TRAIT, THE **GENOTYPE** IS INHERITED **INDEPENDENTLY,** WITH NO ONE TRAIT AFFECTING ANOTHER.

ALL TRAITS HAVE AN **EQUAL** CHANCE OF BEING SORTED OUT AND COMBINED WITH ANY OTHER TRAIT.

BOTANY

AND -- OF **COURSE** -- IT'S OUR OLD FRIEND **MEIOSIS** THAT SETS THE STAGE FOR THESE RECOMBINATIONS!

I DO ALL THE HEAVY LIFTING!

THOSE, THEN, ARE THE **BASIC PRINCIPLES** THAT GOVERN INHERITANCE. MENDEL DISCOVERED THEM, AND THEY HAVE STOOD THE TEST OF TIME PRETTY WELL. TO RECAP...

ALTERNATE VERSIONS OF GENES CALLED **ALLELES** ARE RESPONSIBLE FOR **VARIATIONS** IN ANY SPECIFIC TRAIT--

Every individual inherits **TWO ALLELES** for every trait.

One comes from each parent where they **RANDOMLY SEG-REGATED** from their sister allele during **MEIOSIS.**

ALSO DURING MEIOSIS, EVERY ALLELE IS **SORTED INDEPENDENTLY** OF ALL OTHER ALLELES IN THE PROCESS OF CROSSING-OVER.

EVERY ALLELIC COMBINATION IS POSSIBLE.

AND, if the two alleles inherited by an individual are different, then **ONE** phenotype will always dominate the **OTHER!**

VERY **GOOD**, YOUR GALACTIC ASTUTENESS!

THOSE ARE THE PRINCIPLES THAT HAVE FORMED THE **BEDROCK** OF HUMAN UNDERSTANDING OF GENETICS. THEY MAKE PATTERNS OF INHERITANCE **PREDICTABLE**.

BUT...

...This is where you say **BUT** and let me know that this inheritance business isn't as **CUT-AND-DRIED** as I'd wish it to be...

RIGHT **AGAIN**, MY MOST SAGACIOUS LORDSHIP.

MENDEL'S OBSERVATIONS WERE **SOLID** -- BUT HIS CAREFUL CHOICE OF SUBJECT AND TRAITS INVOLVED **SIMPLE INHERITANCE** ONLY. THAT WAS JUST A **START**.

HUMANS HAVE SINCE LEARNED THAT INHERITANCE ISN'T ALWAYS **SIMPLE**. THERE ARE IMPORTANT EXCEPTIONS TO MENDEL'S OBSERVATIONS.

That figures. There's **NOTHING** simple about genetics...

WELL, IT'S NOT **EASY**, BUT IF YOU GET THE **BASIC CONCEPTS**, THE MORE COMPLEX STUFF ALL MAKES SENSE.

IN **COMPLEX INHERITANCE** CASES, IT'S IMPORTANT TO REALIZE THAT THE EXCEPTIONS TO MENDEL'S PRINCIPLES ALL INVOLVE HOW **PHENOTYPES** -- PHYSICAL TRAITS -- EXPRESS. SEGREGATION AND INDEPENDENT ASSORTMENT, THE LAWS GOVERNING THE INHERITANCE OF **GENOTYPE**, REMAIN **CONSTANT**.

I FIGURED THEY WOULD.

FOR INSTANCE, SIMPLE INHERITANCE DOES NOT ACCOMMODATE THE FACT THAT CERTAIN GENES DO NOT FOLLOW THE STRICT "**DOMINANT MASKS RECESSIVE**" FORM OF EXPRESSION.

THESE GENES EXPRESS IN DIFFERENT **DEGREES** OF DOMINANCE, RESULTING IN A PHENOTYPE SOMEWHERE **BETWEEN** THE DOMINANT AND RECESSIVE VARIATIONS...

AND IN MY DEFENSE I MUST SAY I NEVER CLAIMED THAT MY PRINCIPLES APPLIED TO ANYTHING OTHER THAN MY **PEA PLANTS**.

SNAPDRAGON FLOWERS ARE A GOOD EXAMPLE OF THIS **INCOMPLETE DOMINANCE**. FOLLOWING MENDEL'S EXPERIMENTS, IF WE CROSS A DOMINANT PURPLE-COLORED SNAPDRAGON (PP) WITH A RECESSIVE **WHITE** SNAPDRAGON (pp), WE WOULD **EXPECT** THE F_1 OFFSPRING (Pp) TO ALL EXPRESS THE DOMINANT **PURPLE** COLOR.

BUT, INSTEAD, WHAT WE GET IN SNAPDRAGON F_1 OFFSPRING IS A **LIGHT PURPLE** FLOWER THAT **APPEARS** TO BE THE RESULT OF A **BLEND** OF THE PURPLE AND WHITE COLOR ALLELES!

THIS **INCOMPLETE DOMINANCE** CONTINUES DOWN THE GENERATIONS, WITH THE F_2 POSSIBILITIES BREAKING DOWN IN THE 1:2:1 RATIO -- ¼ **PURPLE** (PP), ½ **LIGHT PURPLE** (Pp OR pP), AND ¼ **WHITE** (pp).

HETEROZYGOUS (Pp, pP) OFFSPRING **ALWAYS** DISPLAY INCOMPLETE DOMINANCE, WITH LIGHT PURPLE FLOWERS THE RESULT.

BUT THE TWO ALLELES ARE **NOT** BLENDING THEIR TRAITS. **GENOTYPICALLY**, THEY CONTINUE TO OBSERVE THE LAWS OF SEGREGATION AND INDEPENDENT ASSORTMENT. THE DISTRIBUTION OF ALLELES DOWN THROUGH THE GENERATIONS REMAINS AS PREDICTABLE AS MENDEL'S FINDINGS.

OF COURSE I OBEY THE LAWS! I JUST LIKE TO **DRESS** MYSELF A BIT DIFFERENTLY...

ONLY THE **PHENOTYPE** DIFFERS, CAUSED BY **CELLULAR MECHANICS** WORKING OUT THE EXPRESSION OF THE COLOR PURPLE IN A DIFFERENT, MORE COMPLEX MANNER.

IT'S NOT HARD TO UNDERSTAND, HOWEVER, HOW PRE-MENDELIAN GENETICS MIGHT HAVE SEEN EXAMPLES OF INCOMPLETE DOMINANCE AS EVIDENCE FOR THE NOTION THAT THE TRAITS OF MOTHER AND FATHER WERE **BLENDED TOGETHER** IN THE OFFSPRING.

I MEAN, HOW **OBVIOUS** CAN THE EVIDENCE FOR HEREDITARY BLENDING BE?

ANOTHER VARIATION IN GENE EXPRESSION INVOLVES **CODOMINANCE**...

...WHICH OCCURS WHEN ALLELES **SHARE EQUALLY** IN THE EXPRESSION OF PHENOTYPE. **BOTH** ALLELES EXPRESS **FULLY**, TOO-- NO COMPROMISES!

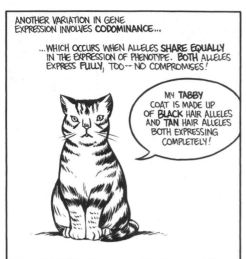

MY **TABBY** COAT IS MADE UP OF **BLACK** HAIR ALLELES AND **TAN** HAIR ALLELES BOTH EXPRESSING COMPLETELY!

HUMAN BLOOD TYPES ARE A VERY WELL-KNOWN EXAMPLE OF CODOMINANCE.

BLOOD COMES IN DIFFERENT TYPES BECAUSE OF PROTEINS CALLED **ANTIGENS** THAT LIE ON THE SURFACE OF BLOOD CELLS, HELPING TO PROTECT HUMANS FROM DISEASE. THERE IS MORE THAN ONE KIND OF ANTIGEN...

...AND SEVERAL DIFFERENT ALLELES ARE RESPONSIBLE FOR PRODUCING THEM. THE MOST WIDESPREAD TYPES ARE CALLED **A** AND **B**-- ALLELES CODING FOR THESE ARE BOTH **DOMINANT**.

WE'RE THE **A** TEAM! WE **DOMINATE**!

WE'RE THE **B** TEAM! WE **DOMINATE**!

BUT SOME HUMANS CARRY TYPE **AB** BLOOD, A SITUATION WHERE **BOTH** ANTIGENS ARE PRESENT BECAUSE BOTH DOMINANT ALLELES HAVE **SIMULTANEOUSLY** AND **FULLY** EXPRESSED!

THE AB GENOTYPE EQUALS AN AB PHENOTYPE!

TOGETHER, WE CAN DOMINATE EVEN **BETTER**!

CODOMINANCE IS ALSO SEEN IN **DOMESTIC CATTLE** THAT SHOW THE REDDISH BROWN COLORATION CALLED **ROAN**.

ROAN OCCURS WHEN THE DOMINANT ALLELE FOR **BROWN HAIR** EXPRESSES **EQUALLY** WITH A DOMINANT ALLELE FOR **WHITE**.

THE RESULT OF THE TWO FULLY EXPRESSED ALLELES IS THE MIX OF HAIR COLORS THAT APPEARS REDDISH BROWN-- **ROAN**.

BUT-- ONCE AGAIN-- THE CODOMINANCE VARIATION IS ALL IN THE PHYSICAL EXPRESSION OF THE TRAITS, THE **PHENOTYPE**. THE **GENOTYPE** STILL FOLLOWS THE RULES WE LEARNED FROM MENDEL.

NOW, SOMETIMES DOMINANT ALLELES DO NOT **EXPRESS** THEIR FULL INFLUENCE **CONSISTENTLY.**

THIS LACK OF TOTAL EXPRESSION IN THE PHENOTYPE IS CALLED **INCOMPLETE PENETRANCE.**

WHEN **100%** OF INDIVIDUALS CARRYING A PARTICULAR DOMINANT ALLELE EXPRESS THE PHENOTYPE, THEN THE ALLELE IS SAID TO HAVE **COMPLETE PENETRANCE.**

MOST DOMINANT ALLELES **ARE** COMPLETELY PENETRANT.

INCOMPLETE PENETRANCE, HOWEVER, IS NOT UNCOMMON. AND A GOOD THING, TOO, IN THE CASE OF THIS EXAMPLE:

BREAST CANCER IS AN ALL-TOO-WIDESPREAD HUMAN SCOURGE THAT CAN BEGIN AS A **MUTATION** -- A SPONTANEOUS GENETIC CHANGE-- ARISING IN ONE OF A NUMBER OF DIFFERENT GENES AND THEN BE TRANSMITTED TO FUTURE GENERATIONS AS AN AUTOSOMAL DOMINANT TRAIT.

I'M **ONE MUTATED CELL** AND I'M JUST THE START!

THOSE MUTATED GENES-- DOMINANT ALLELES-- TEND TO DISPLAY INCOMPLETE PENETRANCE, HOWEVER.

ONLY A **PERCENTAGE** OF HUMANS CARRYING THIS POTENTIAL FOR BREAST CANCER IN THEIR GENOTYPE WILL ACTUALLY DEVELOP THE DISEASE ITSELF...

...ESTIMATES VARY BETWEEN 50 AND 85%.

ANOTHER, SLIGHTLY WEIRD EXAMPLE OF INCOMPLETE PENETRANCE: HUMANS ARE **BILATERAL** CREATURES NORMALLY CARRYING A COMPLEMENT OF **FIVE DIGITS** AT THE EXTREME ENDS OF EACH OF THEIR LIMBS. THEY CALL THEM FINGERS AND TOES.

IN **CERTAIN** HUMAN POPULATIONS, A TENDENCY TOWARD EXPRESSING **MORE** THAN FIVE DIGITS PER LIMB IS CARRIED IN THE GENOTYPE.

THIS IS, INTERESTINGLY ENOUGH, A **DOMINANT** TRAIT...

...WHICH, THANKFULLY FOR THOSE POPULATIONS, EXPRESSES WITH VERY **INCOMPLETE PENETRANCE.**

As we've already discussed, Mendel was **LUCKY**--OR **CLEVER**-- enough to study traits that expressed in only **TWO** variations.

I prefer to think of myself as **INSIGHTFUL**.

But the situation **CAN** become much more complicated.

Alleles can come in **MORE** than just two variations...

...although any individual diploid organism can, of course, hold only **TWO** of those variations in its particular genotype.

Alleles--variants of genes-- are, apparently, the results of mutations and are generally **BENEFICIAL** to the organism.

There's no reason why there would be a **LIMIT** on the number of alleles that might possibly develop from any given gene.

A very common example of the interaction of multiple alleles is a small, furry domesticate called the **RABBIT**.

Its wide variety of **FUR COAT COLORS** comes courtesy of **FOUR ALLELES** that interact and form an arrangement called a **DOMINANCE HIERARCHY**. Here's how it works...

The color **BROWN** (C⁺) is considered the default, "normal" coloration-- which is often referred to as **WILD TYPE**, on the assumption that this was the coloration before domestication and manipulative breeding.

My **BROWN** allele is **DOMINANT** over the **OTHER** three hair-color alleles.

The recessive **ALBINO** gene (c) must be **HOMOZYGOUS**-- carry identical alleles-- to express. It codes for the total **ABSENCE** of color, resulting in white fur, and pink eyes and skin.

Because this trait is actually the **LACK** of a trait (no color), albinism is the result of what's known as a **NON-FUNCTIONING** allele.

THE **CHINCHILLA** ALLELE (C^ch) EXPRESSES AS A **SOLID GRAY** FUR COAT...

...AND IS **PARTIALLY DOMINANT** OVER **HIMALAYAN**!

THE FOURTH ALLELE-- **HIMALAYAN** (C^h)-- SHOWS AS **WHITE** FUR WITH DARK EARS, NOSE, AND **FEET**.

CHINCHILLA / HIMALAYAN CROSSES ARE **HETEROZYGOUS** AND EXPRESS AS GRAY FUR WITH DARK EARS, NOSE, AND FEET!

RABBIT FUR COLOR IS **ACTUALLY** DETERMINED BY **FIVE DIFFERENT GENES.** ALL THAT GENE INTERACTION, MULTIPLIED BY FOUR ALLELES, MAKES FOR A VERY COMPLICATED PICTURE!

BUT-- AS ALWAYS -- MENDEL'S BASIC LAWS OF **SEGREGATION** AND **INDEPENDENT ASSORTMENT** STILL RULE THE GENOTYPE.

BY THE WAY-- THERE ARE SOME ALLELES THAT CARRY A DECIDEDLY **UNBENEFICIAL** TRAIT. LETHAL ALLELES ARE **RECESSIVES** THAT, WHEN HOMOZYGOUS, ALWAYS RESULT IN **DEATH**, OFTEN BEFORE BIRTH.

THERE ARE A FEW EXCEPTIONS: **HUNTINGTON'S DISEASE**, FOR INSTANCE, IS AN **AUTOSOMAL DOMINANT** LETHAL DISORDER THAT DOES NOT EXPRESS UNTIL HUMANS REACH **ADULTHOOD**.

EVEN SO, IT IS ALWAYS ULTIMATELY **FATAL**: 100% PENETRANCE.

the POLITICS of GENETICS

MENDEL'S GENETIC DISCOVERIES TOOK **TIME** AND **COLLABORATIVE EVIDENCE** TO BECOME ACCEPTED AS FACT THROUGHOUT THE SCIENTIFIC COMMUNITY, ALTHOUGH THERE WAS ONE PARTICULARLY NOTEWORTHY **EXCEPTION** TO THE RECOGNITION.

FROM THE 1930s THROUGH THE EARLY '60s, THE TOTALITARIAN RULERS OF THE **UNION OF SOVIET SOCIALIST REPUBLICS** SANCTIONED AN **ALTERNATE** VERSION OF GENETICS CALLED **MICHURINISM**, WHILE CENSURING ALL DISSENTING VIEWPOINTS.

MICHURINISM STOOD IN TOTAL OPPOSITION TO MENDELIAN GENETICS IN ITS STATED BELIEF THAT **PHYSICAL TRAITS** COULD BE DEVELOPED BY **INDIVIDUALS** IN RESPONSE TO THEIR ENVIRONMENT, AND THAT THESE NEWLY DEVELOPED TRAITS COULD THEN BE **TRANS-MITTED** TO THEIR OFFSPRING. (MICHURINISM WAS ACTUALLY A SLIGHT VARIATION ON A PRE-MENDEL THEORY OF HEREDITY AND EVOLUTION CALLED **LAMARCKISM**.)

MICHURINISM WAS PROMOTED TO THE SOVIET LEADERS BY **TROFIM LYSENKO**, AN AGRONOMIST WITH NO SCIENTIFIC TRAINING BUT PLENTY OF MARKETING SAVVY. HE CLAIMED THAT HE HAD PROVEN THE VERACITY OF THE THEORY WITH A PROCESS HE HAD DE-VELOPED CALLED "**VERNALIZATION**," WHICH ALLOWED CEREAL SEEDS THAT WERE PROPERLY **EXPOSED** TO **COLD** AND **MOISTURE** TO **GERMINATE** EARLIER IN THE YEAR THAN THEY OTHERWISE WOULD -- A BENEFICIAL TRAIT THE GRAIN WOULD THEN SUPPOSEDLY PASS ON TO FUTURE GENERATIONS.

IN THE 1930s, THE SOVIET UNION WAS DEALING WITH **FARMER REBELLION** IN THE FACE OF THE UNPOPULAR **COLLECTIVIZATION** OF SOVIET AGRICULTURE. LYSENKO'S "VERNALIZATION," ALTHOUGH UNSUPPORTED BY EVEN A **SHRED** OF SCIENTIFIC EVIDENCE, WAS EXPLOITED BY SOVIET **PROPAGANDISTS** AS AN EFFECTIVE **POPULIST MOTIVATOR:** HANDS-ON "VERNALIZATION" ALLOWED SOVIET FARMERS TO FEEL **EMPOWERED** AND **ENGAGED** BY THE SYSTEM.

ALTHOUGH IT SOON BECAME CLEAR THAT HIS PROCESS PRODUCED **NO** CONSISTENT CROP RESULTS, LYSENKO, A MASTER SELF-PROMOTER, HAD MOVED ON TO **OTHER** SENSATIONALISTIC AGRICULTURAL SCHEMES BASED ON MICHURINISM. NONE WORKED, BUT AS THEY CONTINUED TO SERVE IMPORTANT POLITICAL AND SOCIAL AGENDAS, LYSENKO GAINED **POWER** AND CRUSHED ANY OPPOSING VOICES FROM WITHIN THE SOVIET SCIENTIFIC COMMUNITY.

THE ULTIMATE RESULT WAS THE **ABJECT FAILURE** OF SOVIET AGRICULTURE AS WELL AS THE **DESTRUCTION** OR **RETARDATION** OF ALL **LEGITIMATE** GENETIC INQUIRY AND ACHIEVEMENT.

LYSENKO'S STATE-SANCTIONED PROGRAM IS AN OBJECT LESSON IN WHAT HAPPENS WHEN WHAT IS **WISHED** TO BE TRUE TRUMPS WHAT THE EVIDENCE **DEMONSTRATES** TO BE TRUE. BUT THIS CATASTROPHIC REFUSAL TO ACCEPT FACTS THAT WERE INCONVENIENT OCCURRED LONG AGO, IN A **TOTALITARIAN** SOCIETY, AND COULD **NEVER** HAPPEN IN A **DEMOCRACY**, IN A SCIENTIFICALLY ENLIGHTENED ERA.

OR COULD IT?

THE SUBJECT OF **LETHAL ALLELES** CAN SERVE TO LEAD US INTO ANOTHER GENETIC PHENOMENON CALLED **EPISTASIS**...

ALWAYS GLAD TO BE OF ASSISTANCE...

...WHICH OCCURS WHEN CERTAIN GENES **MASK THE EXPRESSION** OF OTHER GENES.

OUR EXAMPLE ORGANISM WILL BE YET **ANOTHER** DOMESTICATED MAMMAL -- THE **HORSE**. HOW ITS HAIR COLOR EXPRESSES IS A GOOD EXAMPLE OF **EPISTASIS** AS WELL AS OF COMPLEX GENE INTERACTION.

ACTUALLY, THE GENETIC DETERMINATION OF HORSE HAIR COLOR IS **SO** COMPLICATED THAT OUR EXAMPLE HERE HAS BEEN **GREATLY** SIMPLIFIED FROM ALL THE INTER-ACTION THAT **REALLY** HAPPENS.

HORSES CARRY A DOMINANT ALLELE FOR WHITE (**W**) HAIR COLOR.

AS WITH RABBITS, **W** IS ACTUALLY A NONFUNCTIONING ALLELE FOR THE **ABSENCE** OF COLOR. BUT IN ITS **HOMOZYGOUS** (**WW**) FORM IT IS ALSO **LETHAL**.

THERE ARE **NO** HORSES WITH THE **WW** GENOTYPE.

HORSES WITH THE **HETEROZYGOUS Ww** GENOTYPE HAVE A **WHITE** COAT.

ONLY THE **RECESSIVE ww** GENOTYPE ALLOWS **OTHER** GENES TO COME INTO PLAY AND VARIOUS COLORS TO BE EXPRESSED.

THERE ARE ACTUALLY AT LEAST **SEVEN** GENES THAT INTERACT, DETERMINING THE MANY NUANCES OF COAT COLOR SEEN IN HORSES CARRYING THE **ww** GENOTYPE...

...MAKING HORSE BREEDING A KIND OF **FINE ART**!

I'M A HORSE OF A DIFFERENT COLOR!

BUT-- A **DOMINANT W** IN THE GENOTYPE **HIDES** THE ACTIONS OF **ALL** THE OTHER GENES RESPONSIBLE FOR HAIR COLORATION.

THIS IS **DOMINANT EPISTASIS**.

I'VE GOT **LOTS OF COLOR** IN ME...

...BUT I CAN'T GET IT OUT!

THERE ARE QUITE A FEW **MORE** VARIATIONS ON MENDEL'S OBSERVATIONS THAT LAYER EVEN MORE **COMPLEXITY** INTO GENE EXPRESSION.

GENES THAT ARE LOCATED CLOSE TO ONE ANOTHER ON THE CHROMOSOME ARE ALMOST ALWAYS **LINKED**-- INHERITED TOGETHER WITHOUT BENEFIT OF INDEPENDENT ASSORTMENT!

ONE GENE CAN HAVE A HAND IN **MANY** EXPRESSED TRAITS! THIS IS CALLED **PLEIOTROPY**, AND THOSE GENES ARE SAID TO BE PLEIOTROPIC.

AGING MAY BE A RESULT OF **ANTAGONISTIC PLEIOTROPY** -- GENES CREATING EFFECTS THAT **BENEFIT** US WHEN WE'RE YOUNG MAY CREATE **OTHER** EFFECTS THAT **AGE** US LATER IN LIFE!

THERE ARE TRAITS THAT **GROW STRONGER** AS THEY EXPRESS THROUGH SUCCESSIVE GENERATIONS! HUMANS ARE STILL TRYING TO FIGURE OUT THE MECHANICS OF THIS **GENETIC ANTICIPATION!**

THEY KEEP GETTING TALLER!

PARENT

F₁

F₂

AND WHILE MENDEL'S PEA PLANT TRAITS ALL CAME FROM GENES FOUND ON **AUTOSOMAL CHROMOSOMES**, THERE ARE PLENTY OF OTHER GENES ON THE **SEX CHROMOSOMES** -- MOSTLY ON THE **X** CHROMOSOME, ACTUALLY-- THAT EXPRESS DIFFERENTLY DEPENDING ON THE **GENDER** OF THE CARRIER!

THOMAS HUNT MORGAN DISCOVERED THAT FRUIT FLY **EYE** COLOR DIFFERENCES ARE CONTROLLED BY GENES ON THE **Y** CHROMOSOME! CHECK IT OUT!

WHEW! I COULD GO ON...

...BUT THAT'S **ENOUGH** FOR NOW-- I GET THE **POINT**...

...Mendel's law of inheritance will always prove a strong foundation for understanding genotype-- but phenotype expression can be a bit more... UNRULY.

A GOOD ENOUGH SUMMATION, YOUR MOST PATIENT SCHOLARLINESS.

NOW LET'S TAKE A **SHORT BREAK** BEFORE WE CONTINUE WITH A LOOK AT WHAT HUMANS ARE ACTUALLY **DOING** WITH THEIR GROWING KNOWLEDGE OF GENETICS!

CHAPTER 4

Applying All That Stuff-- For the Greater Good

NOW THAT WE'VE COVERED THE BASIC MECHANICS OF GENETICS...

... IT'S TIME WE LOOKED AT THE **PRACTICAL**-- AND SOMETIMES **IMPRACTICAL** -- **APPLICATIONS** TO WHICH HUMANS HAVE PUT THEIR HARD-WON KNOWLEDGE.

LONG BEFORE **MENDEL**, EVEN THE AVERAGE ANIMAL BREEDER UNDERSTOOD THE BASICS OF **PRACTICAL HEREDITY**. STOCK ANIMALS WITH A SPECIFIC TRAIT BRED WITH OTHERS CARRYING THE SAME TRAIT INCREASED THE LIKELIHOOD THAT OFFSPRING WOULD INHERIT THAT TRAIT -- WHETHER BENEFICIAL OR HARMFUL.

AND IN SOME CASES, IT BECAME UNDERSTOOD THAT THE INTER-BREEDING OF STOCK **TOO CLOSELY RELATED BY FAMILY** TENDED TO **INCREASE** THE APPEARANCES OF **UNHEALTHY** TRAITS.

IT WAS SURELY NOTICED THAT SIMILAR UNHAPPY INHERITANCES COULD ALSO BEFALL **HUMAN** POPULATIONS.

HEMOPHILIA AND POLYDACTYLY ARE THE PRICES WE HAVE PAID TO REMAIN BLUE BLOODS!

BUT IT TOOK MENDEL AND HIS **LAWS OF HEREDITY** TO BEGIN THE **SCIENTIFIC, SYSTEMATIC** UNDERSTANDING OF EXACTLY **HOW** AND **WHEN** GENETIC DISORDERS ARE TRANSMITTED...

...AND, WITH THAT, THE HOPE THAT SOMETHING COULD BE DONE TO HELP **GUIDE** INDIVIDUALS PAST GENETIC DANGERS...

WELL, ACTUALLY, *I* WAS ONLY THINKING IN TERMS OF PEA PLANTS.

...SOMETHING THAT BECAME KNOWN AS **GENETIC COUNSELING**...

...THE SYSTEMATIC **EXAMINATION** OF FAMILY HISTORY FOR HEREDITARY DISORDERS AND THE **DETERMINATION** OF THE LIKELIHOOD OF A SPECIFIC FAMILY MEMBER INHERITING A DISORDER, COUPLED WITH PRACTICAL **ADVICE** CONCERNING THE DETERMINATION.

A LITTLE TOO LATE TO HELP **US**.

INDIVIDUALS SEEK THE SERVICES OF **GENETIC COUNSELORS** IF THEY ARE CONCERNED THAT THEY, OR THEIR **OFFSPRING**, MAY BE AT RISK OF A **HERITABLE DISORDER**. A TRAINED COUNSELOR CAN **IDENTIFY** THE **RISKS** AND OFFER **ADVICE** AS TO THEIR NATURE, AS WELL AS **OPTIONS** FOR PREVENTING OR TREATING A SPECIFIC DISORDER.

THE COUNSELOR BEGINS ANY CASE WITH A **SEARCH** OF THE **FAMILY GENETIC HISTORY** OF THE INDIVIDUAL CONSIDERED AT RISK. THAT INDIVIDUAL IS CALLED **THE PROBAND**.

THE SEARCH MAY BE INITIATED BY THE PROBAND, OR BY THE PROBAND'S CONCERNED PARENTS.

THE FAMILY GENETIC HISTORY IS USED TO CONSTRUCT A **PEDIGREE** -- A FAMILY TREE THAT CHARTS ALL KNOWN AND PERTINENT FAMILIAL TRAITS.

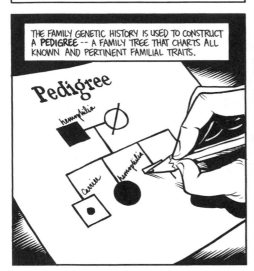

PEDIGREES FEATURE **STANDARDIZED SYMBOLS** TO REPRESENT FAMILY-MEMBER **STATUS**, RELATIVE **RELATIONSHIP**, AND HEALTH ISSUES.

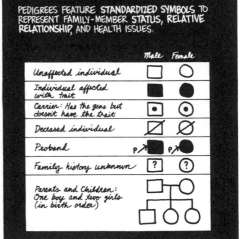

	Male	Female
Unaffected individual	□	○
Individual affected with trait	■	●
Carrier: Has the gene but doesn't have the trait	⊡	⊙
Deceased individual	⊘	⊘
Proband	P↗ ■	P↗ ●
Family history unknown	?	?
Parents and Children: One boy and two girls (in birth order)		

THE COUNSELOR OBTAINS THE BEST FAMILY MEDICAL HISTORY POSSIBLE FROM THE **PROBAND**, OR THE PROBAND'S **PARENTS**, CHARTING DISEASES AND OTHER HEALTH ISSUES AS THEY APPLY TO SPECIFIC FAMILY MEMBERS.

AMONG THE MOST COMMON MEDICAL PROBLEMS LISTED ARE **CANCERS**, **HEART DISEASE**, **ALCOHOLISM**, **DIABETES**, AND **HIGH BLOOD PRESSURE**, AS WELL AS SPECIFIC BIRTH DEFECTS.

AS SPECIFIC GENETIC TRAITS ARE IDENTIFIED WITHIN THE FAMILY...

...THE COUNSELOR'S EXPERTISE WITH MENDELIAN GENETICS HELPS DETERMINE THE CHANCES THAT THE **PROBAND**--

--OR ANY OTHER CONCERNED FAMILY MEMBER--

--MIGHT INHERIT THE TRAIT.

GENETIC COUNSELORS REFER TO ANY FAMILY MEMBER CARRYING **ONE COPY** OF THE GENE CODING FOR A SPECIFIC DISORDER AS **HETEROZYGOUS**...

YOU'RE GOING TO MAKE AN EXAMPLE OF OUR SON AGAIN, AREN'T YOU?

...WHILE ANY MEMBER POSSESSING **TWO COPIES** OF THE GENE CODING FOR A SPECIFIC DISORDER IS CALLED **HOMOZYGOUS**.

SOUND FAMILIAR? IT SHOULD-- IT'S THE **SAME** STANDARD MENDELIAN TERMINOLOGY THAT APPLIES TO PEA PLANTS.

HIS MOTHER AND I DIDN'T EXPRESS, BUT...

ANY FAMILY MEMBER **EXPRESSING** A SPECIFIC DISORDER IS TERMED **AFFECTED**.

ALTHOUGH OUR **POLYDACTYLY** IS A DOMINANT DISORDER, MY PARENTS ARE **NOT** AFFECTED BECAUSE IT IS ALSO **INCOMPLETELY PENETRANT** AND HAS NOT EXPRESSED IN THEM.

AND AS WE'VE LEARNED, DIFFERENT TRAITS, DISORDERS **INCLUDED**, ARE TRANSMITTED GENERATION TO GENERATION IN DIFFERENT PATTERNS-- **MODES OF INHERITANCE** -- DEPENDING ON DOMINANT AND RECESSIVE CHARACTERISTICS.

PARENT

F_1

F_2

HERITABLE TRAITS IN HUMANS FOLLOW THE SAME RULES OF TRANSMISSION AS DISCOVERED BY MENDEL-- WITH A FEW **CURVES** THROWN IN BY GENES LOCATED ON THE **SEX CHROMOSOMES**.

REMEMBER--

--FEMALES CARRY **TWO X CHROMOSOMES** AND **NO Y**, WHILE MALES HAVE A Y AND **ONLY ONE X**.

MODES of INHERITANCE
CHECKLIST

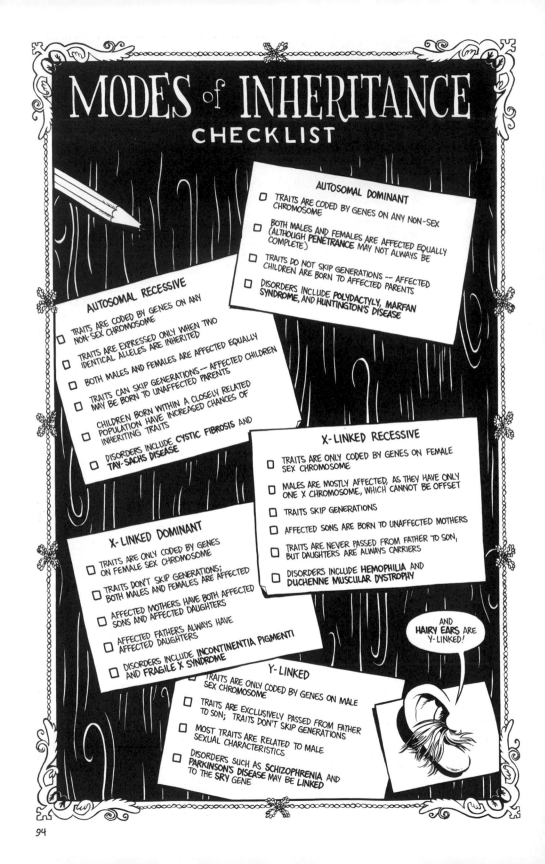

AUTOSOMAL DOMINANT

- ☐ TRAITS ARE CODED BY GENES ON ANY NON-SEX CHROMOSOME
- ☐ BOTH MALES AND FEMALES ARE AFFECTED EQUALLY (ALTHOUGH **PENETRANCE** MAY NOT ALWAYS BE COMPLETE)
- ☐ TRAITS DO NOT SKIP GENERATIONS -- AFFECTED CHILDREN ARE BORN TO AFFECTED PARENTS
- ☐ DISORDERS INCLUDE **POLYDACTLY, MARFAN SYNDROME, AND HUNTINGTON'S DISEASE**

AUTOSOMAL RECESSIVE

- ☐ TRAITS ARE CODED BY GENES ON ANY NON-SEX CHROMOSOME
- ☐ TRAITS ARE EXPRESSED ONLY WHEN TWO IDENTICAL ALLELES ARE INHERITED
- ☐ BOTH MALES AND FEMALES ARE AFFECTED EQUALLY
- ☐ TRAITS CAN SKIP GENERATIONS-- AFFECTED CHILDREN MAY BE BORN TO UNAFFECTED PARENTS
- ☐ CHILDREN BORN WITHIN A CLOSELY RELATED POPULATION HAVE INCREASED CHANCES OF INHERITING TRAITS
- ☐ DISORDERS INCLUDE **CYSTIC FIBROSIS** AND **TAY-SACHS DISEASE**

X-LINKED RECESSIVE

- ☐ TRAITS ARE ONLY CODED BY GENES ON FEMALE SEX CHROMOSOME
- ☐ MALES ARE MOSTLY AFFECTED, AS THEY HAVE ONLY ONE X CHROMOSOME, WHICH CANNOT BE OFFSET
- ☐ TRAITS SKIP GENERATIONS
- ☐ AFFECTED SONS ARE BORN TO UNAFFECTED MOTHERS
- ☐ TRAITS ARE NEVER PASSED FROM FATHER TO SON, BUT DAUGHTERS ARE ALWAYS CARRIERS
- ☐ DISORDERS INCLUDE **HEMOPHILIA** AND **DUCHENNE MUSCULAR DYSTROPHY**

X-LINKED DOMINANT

- ☐ TRAITS ARE ONLY CODED BY GENES ON FEMALE SEX CHROMOSOME
- ☐ TRAITS DON'T SKIP GENERATIONS; BOTH MALES AND FEMALES ARE AFFECTED
- ☐ AFFECTED MOTHERS HAVE BOTH AFFECTED SONS AND AFFECTED DAUGHTERS
- ☐ AFFECTED FATHERS ALWAYS HAVE AFFECTED DAUGHTERS
- ☐ DISORDERS INCLUDE **INCONTINENTIA PIGMENTI** AND **FRAGILE X SYNDROME**

AND **HAIRY EARS** ARE Y-LINKED!

Y-LINKED

- ☐ TRAITS ARE ONLY CODED BY GENES ON MALE SEX CHROMOSOME
- ☐ TRAITS ARE EXCLUSIVELY PASSED FROM FATHER TO SON; TRAITS DON'T SKIP GENERATIONS
- ☐ MOST TRAITS ARE RELATED TO MALE SEXUAL CHARACTERISTICS
- ☐ DISORDERS SUCH AS **SCHIZOPHRENIA** AND **PARKINSON'S DISEASE** MAY BE **LINKED** TO THE **SRY** GENE

THOSE, THEN, ARE THE MODES BY WHICH TRAITS ARE PASSED FROM PARENT TO OFFSPRING. GENETIC COUNSELORS CAN FOLLOW THE PROBABLE PATHS OF ANY DISORDERS FOUND WITHIN A FAMILY AND SO DETERMINE THE CHANCES OF THE PROBAND BEING AFFECTED.

IF THE PROBAND IS DETERMINED TO BE AT RISK, THE COUNSELOR MAY THEN ADVISE **GENETIC TESTING** TO **VERIFY** THE DISORDER.

INDIVIDUALS WHO MAY BE **PARTICULARLY** INTERESTED IN CONSULTING A GENETIC COUNSELOR INCLUDE:

- ANYONE WITH A KNOWN OR SUSPECTED **FAMILY HISTORY** OF A DISORDER OR DISEASE

- ANYONE FROM A **POPULATION** WITH A **HISTORICALLY HIGH RISK** OF A SPECIFIC DISORDER

- PREGNANT WOMEN OVER **35** YEARS OF AGE

- **ANYONE** PLANNING TO HAVE CHILDREN

IF A GENETIC DISORDER HAS BEEN CONFIRMED OR DETERMINED TO BE LIKELY, A GENETIC COUNSELOR CAN PROVIDE BOTH **PSYCHOLOGICAL** ASSISTANCE AS WELL AS GUIDANCE IN SEEKING **THERAPEUTIC RESOURCES**...

...WHICH BRINGS US FROM **DIAGNOSIS** TO THE POTENTIAL FOR **TREATMENT** OF GENETIC DISORDERS AND DISEASES.

IT'S ONLY IN A POST-MENDELIAN WORLD THAT CAN READ THE **DNA** WITHIN THE GENE THAT THIS LEVEL OF UNDERSTANDING AND TREATMENT HAS BECOME POSSIBLE.

BUT MANIPULATING **ANYTHING** ON A GENETIC LEVEL MUST BE BASED ON A KNOWLEDGE OF WHICH GENES EXPRESS WHICH PHENOTYPE, PRECISELY **WHERE** THOSE GENES ARE LOCATED...

... AND THE **EXACT ORDER** OF EVERY ONE OF THE **3 BILLION** NUCLEOTIDE BASE PAIRS CODING THE HUMAN GENOME. UNLOCKING THIS NECESSARY INFORMATION WAS AN EXTENDED UNDERTAKING THAT BEGAN...

...IN THE EARLY 20th CENTURY WHEN BIOLOGIST **THOMAS HUNT MORGAN** BECAME INTERESTED IN DETERMINING THE OPERATING MECHANISMS **BEHIND** MENDELIAN GENETICS.

TO THAT END, MORGAN AND HIS STUDENTS BEGAN OBSERVING **FRUIT FLIES** FOR **MUTANT VARIATIONS.** THEY CROSSBRED MUTATIONS TO OBSERVE **PATTERNS** OF INHERITANCE.

THERE'S A GOOD BUZZ SURROUNDING OUR RESEARCH.

THEY WERE ABLE TO IDENTIFY SOME **COMPLEX** GENE TRANSMISSIONS -- INCLUDING THOSE INVOLVING GENES LOCATED ON THE SEX CHROMOSOMES -- AND RECOGNIZED THAT **LINKAGES** COULD OCCUR BETWEEN GENES LOCATED CLOSELY TOGETHER.

THIS LED TO THEIR BREAKTHROUGH REALIZATION THAT **GENE LOCATIONS** COULD BE DETERMINED WITHIN THE CHROMOSOMES!

WE'VE BRED THIS ONE TO BE **VERY** COMPLEX!

ALFRED STURTEVANT, ONE OF MORGAN'S RESEARCHERS, DISCOVERED THE PRINCIPLES UNDERLYING **GENE MAPPING.** HE WAS ABLE TO MANAGE AND INTERPRET HIS ACCUMULATING GENETIC DATA BY CREATING **SCHEMATIC DRAWINGS** POSITIONING FRUIT FLY GENES IN RELATION TO ONE ANOTHER ON THEIR CHROMOSOMES.

AND SO THE FIRST UNDERSTANDING OF THE **PHYSICAL ORGANIZATION** OF GENES HAD BEGUN --

-- THE FIRST STEPS TOWARD ASSOCIATING SPECIFIC TRAITS WITH SPECIFIC GENES FOUND AT SPECIFIC CHROMOSOMAL LOCATIONS.

FIFTY YEARS LATER, **WATSON** AND **CRICK** CRACKED OPEN THE DNA MACROMOLECULES OF WHICH CHROMOSOMES ARE BUILT...

...AND IDENTIFIED THE SEQUENTIAL ARRANGEMENTS OF THE FOUR BASES **A, T, C,** AND **G** AS THE **KEYS** TO READING THE **INSTRUCTIONS** ENCODED IN GENES...

...OPENING UP A NEW **WORLD** OF POSSIBILITIES. THE **LANGUAGE** CONTROLLING GENE EXPRESSION HAD BEEN DISCOVERED. IT WAS NOW WITHIN THE REALM OF POSSIBILITY THAT HUMANS MIGHT EVENTUALLY BE ABLE TO ALTER THEIR GENETIC FATE.

BUT, TO TURN POSSIBILITY INTO **REALITY**, SCIENTISTS HAD TO LEARN TO **READ** THAT LANGUAGE, TO **INTERPRET** AN ALPHABET COMPOSED OF THE BASE PAIRS C AND G, AND A AND T...

OKAY -- TAKE THIS DOWN -- AATCCGTGATACAGGTAC...

...SEQUENCED IN THE ENDLESS COMBINATIONS THAT MAKE UP EVERY EARTHLY ORGANISM'S INSTRUCTIONS FOR ASSEMBLY AND MAINTENANCE.

CONSIDERING THAT THE HUMAN GENOME -- AN AVERAGE-SIZE GENOME FOR AN EARTHLY ORGANISM -- IS MADE UP OF AT LEAST **3 BILLION BASE PAIRS**, THIS HAD TO HAVE APPEARED AS AN IMPOSSIBLY TIME-CONSUMING TASK.

AY CARAMBA!

THEN, IN THE MID-1970s, BIOCHEMIST **FREDERICK SANGER** DEVELOPED AN EFFICIENT METHOD FOR **SEQUENCING** DNA -- FOR **DETERMINING** THE BASE-PAIR SEQUENCES IN LONG CHAINS OF DNA -- AND THE DOOR SWUNG WIDE OPEN ON THE FUTURE OF GENETICS!

SANGER'S METHOD OF **CHAIN TERMINATION SEQUENCING** PROCESSED STRANDS OF DNA TO IDENTIFY EVERY BASE IN ITS PROPER ORDER -- AND **MUCH** FASTER THAN WAS EVER PREVIOUSLY POSSIBLE.

GENETIC INSTRUCTIONS COULD NOW BE "DECODED" -- AND POTENTIALLY DUPLICATED OR ALTERED!

SANGER'S WORK SET THE STAGE FOR MANY IMPORTANT DEVELOPMENTS, BUT PERHAPS THE MOST IMPORTANT WAS THAT OF A GLOBALLY COOPERATIVE UNDERTAKING OF **STAGGERING** COMPLEXITY AND **ENDLESS** POTENTIAL...

CHEMISTRY · BIOLOGY · PHYSICS · ETHICS · INFORMATICS · ENGINEERING

...THE HUMAN GENOME PROJECT!

THE **HGP** WAS INITIATED IN **1990**, WITH THE EXPRESS **GOAL** OF SEQUENCING ALL 3 BILLION BASE PAIRS IN THE **HUMAN GENOME**...

...BUT A LOT OF **NONHUMAN** GENETIC INFORMATION WOULD ALSO BE GATHERED ALONG THE WAY!

THE OBJECTIVE WAS TO PROVIDE RESEARCHERS AN **UNPRECEDENTED** DEPTH AND BREADTH OF **UNDERSTANDING** OF THE GENETIC FACTORS BEHIND HUMAN DISEASES AND DISORDERS...

...AND THE POTENTIAL FOR CREATING **NEW STRATEGIES** OF DIAGNOSIS, PREVENTION, AND TREATMENT.

THE GOAL WAS ACTUALLY ACCOMPLISHED TWO YEARS **AHEAD** OF SCHEDULE, THANKS IN LARGE PART TO CONTINUAL ADVANCES IN DNA SEQUENCING METHODOLOGY AND RELIABILITY.

IT WAS A **TRULY** AMAZING ACCOMPLISHMENT...

...BY **ANY** INTELLIGENT SPECIES' STANDARDS!

IN **2003** THE HGP CELEBRATED PROJECT COMPLETION. THE HUMAN GENOME HAD BEEN SEQUENCED TO A DEGREE OF **99.99%** ACCURACY.

H.G.P.

THE **INTERNATIONAL STANDARD** SETS THE DEGREE OF ACCEPTABLE ERROR AT ONE FOR EVERY **10,000** BASE PAIRS. 99.99% -- THAT'S PRETTY DARN ACCURATE.

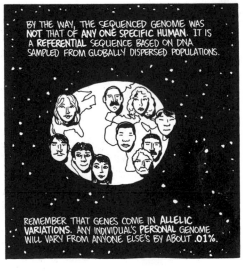

BY THE WAY, THE SEQUENCED GENOME WAS **NOT** THAT OF **ANY ONE SPECIFIC HUMAN.** IT IS A **REFERENTIAL** SEQUENCE BASED ON DNA SAMPLED FROM GLOBALLY DISPERSED POPULATIONS.

REMEMBER THAT GENES COME IN **ALLELIC** VARIATIONS. ANY INDIVIDUAL'S **PERSONAL** GENOME WILL VARY FROM ANYONE ELSE'S BY ABOUT **.01%**.

BUT **BEFORE** THE HGP COMPLETED THE **HUMAN** GENOME, THEY DECIPHERED THE SEQUENCING OF SOME MUCH SIMPLER ORGANISMS.

MANY GENES ARE PRACTICALLY **IDENTICAL** FROM SPECIES TO SPECIES -- ESPECIALLY THOSE RESPONSIBLE FOR META- BOLIC PROCESSES. LEARNING THE BASE SEQUENCES THAT COMPRISE THE GENES OF **ANY** ORGANISM IS LIKELY TO YIELD PLENTY OF INFORMATION ABOUT **OTHER** ORGANISMS.

YOU AND I, WE HAVE A LOT IN COMMON.

RESEARCHERS DID NOT HAVE TO START FROM **SCRATCH** TO DECIPHER THE BASE SEQUENCES OF THE APPROXIMATELY 25,000 HUMAN GENES. **MANY** HAD BEEN SEQUENCED WHILE WORKING ON THE GENOMES OF OTHER, LESS COMPLEX ORGANISMS.

IT SEEMS WE SHARE A GENE FOR **CHEESE** APPRECIATION.

THE VERY **FIRST** GENOME COMPLETELY SEQUENCED -- IN **1995** -- WAS THAT OF A STRAIN OF **BACTERIA**. IT HAS A RELATIVELY MANAGEABLE COMPLEMENT OF 1.9 MILLION BASE PAIRS AND 1,700 GENES.

WE'RE ALSO THE MOST **SUCCESSFUL** ORGANISMS ON EARTH! KEEP IT **SIMPLE**, STUPID!

BREWER'S YEAST WAS THE FIRST **EUKARYOTE** SEQUENCED. THIS FUNGUS IS AN ORGANISM CON- SIDERABLY MORE CLOSELY RELATED TO HUMANS, WITH **30%** OF ITS GENES **ANALOGOUS** TO HUMAN GENES.

WE SHARE A COMMON INTEREST, **TOO...**

BREWER'S YEAST IS IMPORTANT IN **FOOD PRODUCTION** -- IT MAKES BREAD **RISE** AND **FERMENTS** BEER AND WINE.

SEQUENCING ITS GENOME PROVIDED VALUABLE INFORMATION TO **AGRICULTURAL SCIENCE...**

hic

99

...WHILE ALSO PROVIDING IMPORTANT INSIGHTS INTO HOW GENES **CLOSELY SITUATED** ON THEIR CHROMOSOMES WORK **TOGETHER** TO EXPRESS THEIR FUNCTIONS.

BREWER'S YEAST'S GENES GAVE CLUES TO SIMILAR **COMPLEX GENE EXPRESSION** IN HUMANS.

WE'VE **BOTH** HELPED WITH THAT.

THE HUMANS HAVE DISCOVERED THAT INCURABLE AND POTENTIALLY FATAL DISEASES -- SUCH AS **ALZHEIMER'S** AND **DIABETES** -- ARE PASSED FROM ONE GENERATION TO THE NEXT VIA COMPLEX GENE EXPRESSION.

LOTS OF COMPLEX GENE INTERACTION MAKES A COMPLETE UNDERSTANDING OF THESE TERRIBLE AFFLICTIONS DIFFICULT.

You mean to say that the rulers of Earth...

...are learning to decipher their own genetic disorders by studying such **LOWLY ORGANISMS**?!

ABSOLUTELY! REMEMBER-- DNA LINKS **ALL** EARTHLY LIFE.

IT WOULD BE SENSIBLE IF **WE** BEGAN STUDYING **OUR** LOWLY RELATIVE, THE DESPISED **FRIPPURVIAN FLOGWORM,** FOR CLUES TO THE EXPRESSION OF **OUR** COMPLEX GENETIC DISORDER.

AND SPEAKING OF WORMS, THE EARTHLY **ROUNDWORM,** WITH **97 MILLION** BASE PAIRS, WAS THE FIRST **MULTI-CELLULAR** ORGANISM TO HAVE ITS GENOME FULLY SEQUENCED.

ITS **20,000 GENES** ARE ONLY ABOUT 5,000 SHORT OF A HUMAN'S TOTAL. CLEARLY, THE NUMBER OF BASE PAIRS AND GENES INDICATES **LITTLE** ABOUT THE COMPLEXITY OF THE ORGANISM.

WAS THAT SUPPOSED TO BE SOME KIND OF WISECRACK?!

ROUNDWORMS ARE, HOWEVER, AN IMPORTANT STUDY SPECIES IN HUMAN GENE RESEARCH, AS THE TWO ORGANISMS EACH REPRODUCE **SEXUALLY,** HAVE **ANALOGOUS ORGAN AND NERVOUS SYSTEMS,** AND RELATE TO THEIR ENVIRONMENTS THROUGH SHARED **SENSES** SUCH AS SMELL AND TASTE.

YOU AND ME, BABE--WE ARE BOTH VERY **SENSUAL, SEXY** CREATURES...

WITH A FULL **40%** OF THE ROUNDWORM'S GENES EQUIVALENT TO A HUMAN'S, EXPLORATIONS OF ITS GENETIC FUNCTIONS RELATE **DIRECTLY** TO UNDERSTANDING THE HUMAN GENOME.

THE **MALE** HUMAN GENOME, ANYWAY.

WAS THAT SUPPOSED TO BE SOME KIND OF WISECRACK?!

MANY **OTHER** GENOMES WERE SEQUENCED AS WELL ON THE PATH TO DECODING THE **HUMAN** GENOME. THESE GENOMES FOSTER A GREATER UNDERSTANDING OF THE INTERRELATED NATURE OF ALL LIFE ON EARTH.

MOSTLY BACTERIA AND FUNGI AND OTHER MICROBES...

SO WHAT **SPECIFIC** KNOWLEDGE HAS COME FROM THE COMPLETED HUMAN GENOME PROJECT?

PLENTY--

--STARTING WITH A GREATER UNDERSTANDING OF THE GENOME ITSELF.

FOR INSTANCE, IT TURNS OUT THAT ONLY **2%** OF ALL HUMAN DNA COMPRISES THE FUNCTIONING GENES--MEANING GENES THAT ACTUALLY CODE TO CREATE PROTEINS.

NEVER HAVE...

...SO MANY...

...OWED SO MUCH...

...TO SO...

...FEW!

ALL THAT OTHER **98%** ISN'T JUST "JUNK," HOWEVER-- THERE ARE VAST AMOUNTS OF **REPEATING SEQUENCES** OF BASE PAIRS THAT APPARENTLY HAVE PLENTY TO SAY ABOUT **WHEN** AND **HOW OFTEN** GENES EXPRESS.

BUT BECAUSE WE'RE **SO** REPETITIOUS, WE'RE NOT EASY TO DECIPHER!

THERE ARE ALSO REMNANTS OF **HUMAN** --AND POSSIBLY **VIRAL**-- GENETIC HISTORY HIDDEN IN THE GENOME. AT LEAST 37 **NONFUNCTIONING** GENES HAVE BEEN LOCATED. THESE PSEUDOGENES WERE AT ONE TIME IN THE PAST NECESSARY AND ACTIVE. THEY HAVE BEEN **MOTHBALLED**, BUT MAY STILL OFFER IMPORTANT EVOLUTIONARY INFORMATION.

OH, THE STORIES WE COULD TELL...

DISCARDED GENES

THE HUMAN GENOME PROJECT ALSO DISCOVERED THAT AS FUNCTIONING GENES ARE DUPLICATED, THEY MAY GAIN NEW FUNCTIONS.

CHEMISTRY BIOLOGY PHYSICS ETHICS INFORMATICS ENGINEERING

GENOMES ARE DYNAMIC, LIVING SYSTEMS THAT CONSTANTLY EVOLVE!

And I suppose that this **DYNAMIC ACTIVITY** is due to their **SEXUAL NATURE** and all very **GOOD** for the species' health...

WELL, IT'S DUE TO THE GENETIC VARIETY AVAILABLE THROUGH SEXUAL REPRODUCTION **PLUS** THE VARIATIONS THAT OCCUR THROUGH NORMAL MUTATION.

AND WHILE GENOME CHANGES ARE NEGLIGIBLE FROM A **GENERATIONAL** PERSPECTIVE, OVER **TIME** THEY KEEP SPECIES COMPETITIVE AND VITAL.

And **THAT'S** what we **SQUINCH** seem to be missing.

WE CERTAINLY **COULD** USE SOME FRESH GENETIC ALTERNATIVES TO MEET THE DISEASES THAT ARE OVER-TAKING US. BUT, LET ME GET BACK TO THE **HUMAN GENOME**...

THE COMPLETED HUMAN GENOME ANNOUNCED IN 2003 WAS ACTUALLY A **ROUGH DRAFT** OF THE **BASE-PAIR TEXT** FOUND ON EACH CHROMOSOME.

THERE WAS **STILL** MUCH WORK TO BE DONE -- THAT 3-BILLION-BASE-PAIR-LONG CHAIN OF NUCLEOTIDES NEEDED TO BE **INTERPRETED** AS **SPECIFIC GENES, REPEATING SEQUENCES,** AND **EVERYTHING ELSE** THAT IS HELD IN THE GENOME.

FURTHER ADVANCES IN **HIGH-QUALITY SEQUENCING TECHNIQUES** HAVE HELPED FILL IN MUCH OF THE UN-KNOWN TERRITORY ON THE **GENETIC MAP,** AND HAVE PROVIDED **ACCURATE LOCATIONS** FOR THE GENES ON ALL 24 HUMAN CHROMOSOMES.

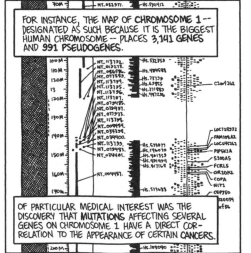

FOR INSTANCE, THE MAP OF **CHROMOSOME 1** -- DESIGNATED AS SUCH BECAUSE IT IS THE BIGGEST HUMAN CHROMOSOME -- PLACES 3,141 GENES AND 991 PSEUDOGENES.

OF PARTICULAR MEDICAL INTEREST WAS THE DISCOVERY THAT **MUTATIONS** AFFECTING SEVERAL GENES ON CHROMOSOME 1 HAVE A DIRECT COR-RELATION TO THE APPEARANCE OF CERTAIN CANCERS.

BUT THE GENES MATCHED WITH **SPECIFIC FUNCTIONS**--OR EVEN **DISORDERS**--REPRESENT JUST A **FRACTION** OF ALL GENES.

= ONE GENE CONTRIBUTING TO EYE COLOR (BLUE)

THERE ARE STILL **HUGE** SWATHS OF THE HUMAN GENOME THAT NEED TO BE INTERPRETED...

...AND HUMAN SCIENTISTS ARE WORKING TO MAKE WHAT **IS** KNOWN OF THE GENOME MORE USER-FRIENDLY. FOR EXAMPLE...

...CANCER RESEARCHERS, IN 2006, BEGAN A MONUMENTAL FOLLOW-UP TO THE HUMAN GENOME PROJECT CALLED...

THE CANCER GENOME ATLAS

...THE CANCER GENOME ATLAS!

CANCERS ARE ABNORMAL, UNCONTROLLED CELL GROWTHS THAT EXACT A TERRIBLE TOLL ON HUMANKIND.

EVEN IN EARTH'S MOST TECHNO-LOGICALLY ADVANCED NATION, THE U.S.A., ONE PERSON DIES FROM CANCER EVERY MINUTE.

STARTING WITH THE VAST AMOUNT OF RAW INFORMATION PROVIDED BY THE **HGP**, THE **GOAL** OF THE ATLAS IS TO EVENTUALLY **IDENTIFY** AND **INDEX** ALL GENES LINKED TO **CANCERS**...

...PROVIDING A POWERFUL, SPECIALIZED TOOL FOR ONGOING RESEARCH AND THERAPY DEVELOPMENT.

THAT WOULD BE A HELP...

THE HUMAN GENOME

LINKING THOSE **GENES** TO THEIR **FUNCTIONS** HAS ALREADY LED TO SOME IMPORTANT DISCOVERIES.

THE GENE DESIGNATED **BRCA 1**, FOR INSTANCE...

BRCA 1 HAS BEEN DEMONSTRATED TO CODE FOR A PROTEIN THAT **NORMALLY** ORCHESTRATES THE **REPAIR** OF DAMAGED DNA. BUT IT ALSO APPEARS TO MODERATE THE EFFECT OF THE GROWTH HORMONE **PROGESTERONE**...

GET THAT G OUT AND THAT C IN -- AND BE QUICK ABOUT IT!

...ON A WOMAN'S **BREAST** AND **OVARY CELLS**. IT HELPS **DESTROY** THE PROTEIN RECEPTORS THAT ALLOW THOSE CELLS TO RECEIVE PROGESTERONE'S MESSAGE...

...ONCE PROPER GROWTH HAS BEEN ATTAINED AND **BEFORE** THE HORMONE CAN URGE **TOO** MUCH UNCONTROLLED GROWTH.

PROGESTERONE

RECEPTOR

OVARY CELL

THAT'S **ENOUGH**, ALREADY! GET **OUTTA** HERE!

PROGESTERONE'S MESSAGE →

MUTATIONS TO BRCA 1, HOWEVER, **DISRUPT** THIS PROCESS, LEAVING THE GENE'S ROLE IN SUPERVISING NORMAL CELL GROWTH **INCOMPLETE** AND SO OPENING THE DOOR TO UNREGULATED AND INVASIVE CELL GROWTH...

...A HALLMARK OF CANCER.

CAN'T REMEMBER WHAT TO DO NEXT...

BRCA 1

MUTATED BRCA 1 GENES ARE JUST **ONE** FACTOR IN THE DEVELOPMENT OF BREAST AND OVARIAN CANCER. BEING ABLE TO **TARGET** BRCA 1 AS WELL AS ALL OTHER CULPRITS HOLDS THE PROMISE OF DEVELOPING THERAPIES THAT MIGHT CORRECT HARMFUL MUTATIONS AND SO BE MORE **EFFECTIVE**, AND LESS **TRAUMATIC**, THAN CURRENT SURGICAL, RADIATION, AND CHEMOTHERAPY TREATMENTS.

GOING FORWARD, **CANCER GENOME ATLAS** RESEARCHERS WILL COMPARE THE DNA OF TUMOR SAMPLES **AGAINST** THE HGP'S REFERENTIAL HUMAN GENOME, EXAMINING GENES SUSPECTED OF ASSOCIATION WITH SPECIFIC CANCERS AND SEARCHING FOR ABNORMALITIES.

The Human GENOME from A to T!

THE CANCER GENOME ATLAS

AND ALTHOUGH PREVAILING THEORY HAS LONG HAD IT THAT **MUTATIONS** IN KEY REGULATORY GENES ARE THE **PRIMARY** CAUSE OF CANCER, IT HAS RECENTLY BEEN SUGGESTED THAT ABNORMALITIES ON A **CHROMOSOMAL** LEVEL MAY BE AS MUCH TO BLAME!

WHAT?!

THERE IS PERHAPS NO ASPECT OF GENETICS SO COMPLETELY MISUNDERSTOOD AS THAT OF THE CONCEPT OF MUTATION. POPULAR CULTURE HAS DONE A POWERFUL JOB OF CREATING A DRAMATIC VISION OF **GIGANTISM**, MONSTROUS EVIL, AND SUPER-POWERED SAVIORS ALL CONVENIENTLY SUMMED UP AS **"MUTANTS."** LOTS OF THRILLS AND CHILLS, BUT MISLEADING.

MISUNDERSTOOD MUTANTS

IN REALITY, MUTATION IS NEITHER INTRINSICALLY GOOD NOR EVIL AND IS CERTAINLY NOT NEARLY SO SENSATION-ALISTIC. IT IS MERELY **NECESSARY.**

MUTATION IS CHANGE ON A GENETIC SCALE. IT IS RESPONSIBLE FOR **ALLELIC VARIATION** WITHIN GENES AND THUS THE CREATION OF **DIVERSITY** BETWEEN ORGANISMS. BECAUSE OF MUTATION, ONE HUMAN MIGHT HAVE BROWN EYES WHILE ANOTHER MIGHT HAVE GREEN. OVER TIME, MUTATIONS ARE RESPONSIBLE FOR **EVOLUTION**--FOR UPGRADING SPECIES, FOR KEEPING THEM COMPETITIVE WITHIN THEIR ENVIRONMENT.

BUT MUTATIONS ARE **NOT** PURPOSE-DRIVEN. THEY ARE **ARBITRARY** CHANGES THAT CAN PRESENT EITHER POSITIVE **OR** NEGATIVE CONSEQUENCES. AT WORST, THEY CAN DISRUPT HEALTHY GENE FUNCTIONS-- **CANCER** IS A RESULT OF MUTATION.

MUTATIONS CAN EITHER OCCUR **SPONTANEOUSLY**--RANDOMLY, WITHOUT EXTERNAL CAUSE-- OR THEY MAY BE **INDUCED** BY EXPOSURE TO RADIATION OR CHEMICAL AGENTS. TOO MUCH ULTRAVIOLET LIGHT, FOR INSTANCE, CAN INDUCE MUTATIONS LEADING TO CANCER.

MUTATION IS, SIMPLY ENOUGH, THE **SUBSTITUTION** OF ONE BASE FOR ANOTHER, OR THE **DELETION** OR **ADDITION** OF A BASE, IN THE LONG CHAIN OF NUCLEOTIDES COMPRISING A GENE. ANY ALTERATIONS TO THE BASE SEQUENCES IN A GENE CAN LEAD TO A CHANGE IN THE **INSTRUCTIONS** READ DURING THE **TRANSLATION** PROCESS THAT BUILDS PROTEINS.

MUTATIONS THAT OCCUR WITHIN **SOMATIC CELLS** -- THE CELLS THAT BUILD AND MAINTAIN AN ORGANISM -- ARE CALLED, NATURALLY ENOUGH, SOMATIC. THOSE THAT APPEAR WITHIN **SEX CELLS** -- SPERM OR EGGS-- ARE KNOWN AS **GERM-CELL MUTATIONS.** SOMATIC MUTATIONS AREN'T HERITABLE -- THEY ONLY AFFECT THE INDIVIDUAL WITH THE MUTATION. GERM-CELL MUTATIONS, HOWEVER, AFFECT THE **OFF-SPRING** OF THE INDIVIDUAL WITH THE ORIGINAL MUTATION AND THEN BECOME HERITABLE BY **SUCCESSIVE GENERATIONS.**

EVERY INDIVIDUAL OF EVERY SPECIES EXPERIENCES COUNTLESS MUTATIONS DURING HIS OR HER LIFETIME (AND THE FREQUENCY INCREASES WITH AGE). **MOST** MUTATIONS HAVE NO APPRECIABLE EFFECT, BUT **SOME** DO AND THAT'S WHY LIFE EVOLVES AND CELLS SOMETIMES TURN AGAINST THEIR NEIGHBORS.

BUT IT'S HIGHLY UNLIKELY THAT PIGS WILL GROW WINGS AND FLY ANYTIME SOON.

KARYOTYPE OF NORMAL HUMAN CHROMOSOME SETS

MANY GENETIC DISORDERS ARE INDISPUTABLY THE RESULT OF **CHROMOSOMAL** ABNORMALITIES--**ANEUPLOIDY** IS A DISORDER IN WHICH THERE IS AN IMBALANCE IN THE NUMBER OF CHROMOSOMES.

REMEMBER--HUMANS ARE NORMALLY **DIPLOID**, MEANING THAT THEY CARRY **TWO SETS** OF THEIR 22 AUTOSOMAL CHROMOSOMES, PLUS TWO SEX CHROMOSOMES (EITHER XX OR XY).

ANEUPLOIDY OCCURS WHEN CHROMOSOMES DO NOT SEG-REGATE PROPERLY DURING **MEIOSIS**, LEAVING EGG OR SPERM CELLS WITH EITHER TOO MANY OR TOO FEW CHROMOSOMES.

CHROMOSOMAL IMBALANCE IS RESPONSIBLE FOR **50%** OF ALL CASES OF MISCARRIAGES...

NONDISJUNCTION

FIRST MEIOTIC DIVISION

SECOND MEIOTIC DIVISION

FERTILIZATION

TRISOMIC
(ONE TOO MANY CHROMOSOMES)

MONOSOMIC
(ONE TOO FEW CHROMOSOMES)

DOWN SYNDROME -- EXTRA CHROMOSOME 21

...AND THOSE OFFSPRING THAT **DO** COME TO TERM ARE AFFLICTED WITH BIRTH DEFECTS. DOWN, CRI DU CHAT, AND KLINEFELTER'S SYNDROMES ARE AMONG THE MANY POSSIBLE DISORDERS.

RECENT ADVANCES IN **DIAGNOSTICS**, AIMED AT DETER-MINING THE STATE OF PLOIDY **BEFORE** PREGNANCY, OFFER THE POSSIBILITY OF GUIDANCE UNAVAILABLE TO EARLIER GENERATIONS.

AND, AS MENTIONED BEFORE, ANEUPLOIDY IS COMMONLY SEEN WITHIN **CANCER** CELLS.

SO, RESEARCH INTO ALL THESE ABNORMAL-ITIES CONTINUES ON MANY FRONTS...

Okay, okay!

I UNDERSTAND-- humans are getting a much better grip on **IDENTIFYING** the genetic sources of many of their maladies...

...but I'm still waiting to hear what they're doing to **CORRECT** their problems!

I WAS JUST **GETTING** THERE, MY MOST IMPATIENT POTENTATE!

HUMAN SCIENTISTS HAVE BEGUN TO WORK **WITHIN** THE GENE, **MANIPULATING** DNA AND EVEN INTRODUCING IT TO NEW LOCATIONS. IT'S ALL ABOUT EXPRESSING THE **RIGHT** PROTEIN IN THE **RIGHT** LOCATION AT THE **RIGHT** TIME.

WELCOME TO THE WILD NEW WORLD OF **GENETIC ENGINEERING!**

IT ALL BEGAN IN 1970, WHEN MICROBIOLOGIST **HAMILTON O. SMITH** IDENTIFIED CERTAIN **RESTRICTION ENZYMES** WITH WHICH HE WAS ABLE TO **SEPARATE** NUCLEOTIDES WITHIN A STRAND OF DNA.

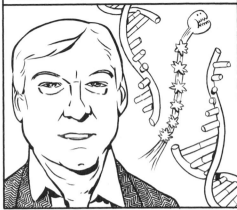

EXPERIMENTING FURTHER, HE DISCOVERED THAT WITH **ANOTHER** ENZYME – **LIGASE** – HE COULD **PASTE** THOSE DISASSEMBLED BITS OF DNA INTO **NEW ARRANGEMENTS.**

THIS TECHNIQUE IS CALLED **GENE SPLICING...**

...OR **RECOMBINANT DNA TECHNOLOGY.**

DNA IS **CUT APART** AND **SPLICED** BACK TOGETHER – **RECOMBINED** – INTO NEW BASE-PAIR SEQUENCES IN ORDER TO CREATE **ARTIFICIAL GENES.**

I FEEL LIKE EXPRESSING MYSELF IN WAYS I NEVER THOUGHT POSSIBLE!

TO WHAT **END**, YOU MIGHT ASK?

MAYBE MOST IMPORTANT, TO PRODUCE PREVIOUSLY IMPOSSIBLE QUANTITIES OF QUALITY-CONTROLLED **PROTEINS** FOR THERAPEUTIC SERVICES.

A MAKE-BELIEVE, GROSSLY SIMPLIFIED EXAMPLE: LET'S SAY THAT THERE IS A GROUP OF HUMANS WHO SUFFER FROM THE LACK OF A PARTICULAR PROTEIN. THEIR BODIES CAN'T **PRODUCE** THAT PROTEIN.

WE WERE BORN WITHOUT THE ABILITY TO SMILE.

THESE HUMANS NEED THIS PROTEIN PROVIDED FOR THEM **AFFORDABLY** AND IN **SUFFICIENT QUANTITY.**

THE GENE THAT PRODUCES THE NEEDED PROTEIN IS, OF COURSE, AVAILABLE IN OTHER HUMANS. BUT IN ORDER TO CREATE SUFFICIENT QUANTITIES OF THAT PROTEIN, THE GENE NEEDS TO BE PLACED IN A **UNIQUE** ENVIRONMENT – A **VECTOR** – THAT WILL ALLOW IT TO REPLICATE **RAPIDLY.**

I MAKE THE SMILE PROTEIN JUST FINE – BUT NOT ENOUGH FOR ALL OF US.

THE PERFECT VECTOR JUST HAPPENS TO LIVE CLOSE BY, IN THE HUMAN GUT. IT'S AN ANCIENT BACTERIAL ALLY NAMED **ESCHERICHIA COLI.**

E. COLI IS THE **PER-FECT** INCUBATOR FOR REPLICATING GENES QUICKLY.

I'VE **ALWAYS** BEEN THERE FOR YOU!

E. COLI

E. COLI IS A **PROKARYOTE** WITH A SIMPLE RING OF DNA CALLED A **PLASMID,** WHICH IS FIRST REMOVED FROM THE BACTERIA.

USING THE RESTRICTION ENZYMES, THE PLASMID IS **CUT UP** AND ITS UNWANTED GENES ARE **REMOVED.** THEN OUR **HUMAN SMILE GENE** IS SPLICED INTO THE RING!

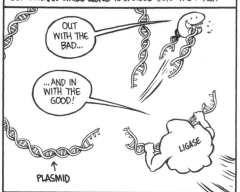

OUT WITH THE BAD...

...AND IN WITH THE GOOD!

LIGASE

↑ PLASMID

THE MODIFIED PLASMID IS PLACED **BACK** IN ITS BACTERIAL HOME AND THE GENETICALLY MANIPULATED E. COLI IS **CULTURED** AND ENCOURAGED TO **MULTIPLY...**

...AND BACTERIA ARE **VERY** GOOD AT MULTIPLYING.

E. COLI REPLICATES ITSELF TO FORM BILLIONS OF COPIES **IN SHORT ORDER.** EVERY COPY OF THE MODIFIED PLASMID WILL CARRY A COPY OF THE INSERTED HUMAN GENE...

...WITH THE MUCH-NEEDED **SMILE PROTEIN** BEING PRODUCED FROM THE DNA INSTRUCTIONS OF **EVERY ONE** OF THOSE BILLIONS AND BILLIONS OF GENES!

WHY DO I FEEL THE URGE TO **SMILE**? I DON'T HAVE A MOUTH...

RECOMBINANT DNA TECHNOLOGY HAS **ALREADY** BEEN USED TO MAKE QUANTITIES OF THESE **REAL** THERAPEUTIC PROTEINS:

INTERFERON

A POWERFUL **VIRUS FIGHTER** WITH GREAT PROMISE. USED TO COMBAT CERTAIN CANCERS.

INSULIN

CRITICAL FOR BREAKING DOWN SUGARS IN HUMAN BLOOD. LACKING IN INDIVIDUALS WITH DIABETES.

SOMATOTROPIN

ALSO KNOWN AS **HUMAN GROWTH HORMONE.** HELPS PREVENT GROWTH DEFICIENCIES, ESPECIALLY IN CHILDREN. AT THIS TIME, THE RECOMBINANT FORM IS STILL BEING TESTED FOR SAFETY.

BUT A **SMILE SERUM** WOULDN'T BE A BAD IDEA, EITHER!

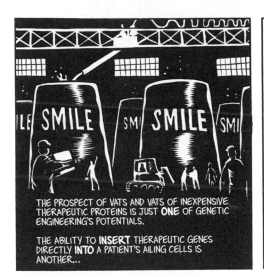

THE PROSPECT OF VATS AND VATS OF INEXPENSIVE THERAPEUTIC PROTEINS IS JUST **ONE** OF GENETIC ENGINEERING'S POTENTIALS.

THE ABILITY TO **INSERT** THERAPEUTIC GENES DIRECTLY **INTO** A PATIENT'S AILING CELLS IS ANOTHER...

...MADE POSSIBLE BY THE USE OF **VIRUSES** AS VECTORS!

VIRUSES ARE BIZARRE THINGS. THEY ARE NEITHER DEAD NOR ALIVE, BUT THEY CARRY THEIR OWN DNA AND THEY **REPRODUCE**-- WHICH MAKES THEM VALUABLE **DELIVERY SYSTEMS.**

INFLUENZA VIRUS

DNA--OR **RNA**--IS REALLY THE **ONLY** THING THAT A VIRUS HAS WITHIN ITS SHELL. TO REPRODUCE, IT MUST CONVERT ITS RNA TO A FORM OF **DNA**, THEN **INVADE** A HOST CELL AND **HIJACK** THAT CELL'S REPLICATION EQUIPMENT.

GIMME ALL YOUR RIBOSOMES AND DON'T ASK QUESTIONS!

SOME VIRUSES USE THE CELL TO KEEP REPLICATING THEMSELVES UNTIL THE CELL IS **BLOWN APART** BY THE POPULATION BOOM! NOT **ALL** ARE SO DESTRUCTIVE TO THEIR HOST, HOWEVER.

THE MOST EFFECTIVE VIRUSES WANT TO KEEP THEIR HOST **ALIVE!**

NOW, THERE IS A **PARTICULAR** TYPE OF VIRUS CALLED A **RETROVIRUS** THAT REPRODUCES BY **SPLICING** ITS NEWLY SYNTHESIZED DNA RIGHT ONTO A HOST CELL'S **CHROMOSOMES.**

SO, AS THE CHROMOSOMES ARE REPLICATED DURING THE NORMAL CELL CYCLE, SO IS THE VIRAL DNA! IT HAS BECOME A **PART** OF ITS HOST'S DNA!

GENE SPLICING IS NOTHING NEW FOR US!

THE **HUMAN IMMUNODEFICIENCY VIRUS** THAT CAUSES **ACQUIRED IMMUNE DEFICIENCY SYNDROME (AIDS)** WORKS THIS WAY -- AND THAT STRATEGY HAS SO FAR KEPT IT INCURABLE.

BUT, ON THE PLUS SIDE, RETROVIRUSES CAN BE MADE TO FIGHT **FOR** HUMAN HEALTH!

HIV RETROVIRUS

WE CAN?

VIRUSES, BASED ON THEIR REPRODUCTIVE STRATEGY, ARE THE PERFECT VECTORS FOR WHAT IS CALLED **GENE THERAPY**-- TREATMENTS THAT EITHER INTRODUCE A **NEW, HEALTHY GENE** INTO A COMPROMISED SYSTEM OR **TURN OFF** GENES RESPONSIBLE FOR UNHEALTHY CELL GROWTH.

WITH THE SAME RECOMBINANT TECHNOLOGY USED TO INSERT GENES INTO BACTERIA, SCIENTISTS HAVE RE-OUTFITTED THESE VIRUSES AS DELIVERY SYSTEMS THAT BEAR A RECOMBINED GENETIC **PAYLOAD** STRAIGHT INTO HUMAN CELLS, WHERE THEY THEN DO WHAT COMES NATURALLY-- **REPRODUCE.**

ONCORETROVIRAL VECTORS, DEVELOPED FROM A VIRUS THAT CAUSES LEUKEMIA IN MONKEYS, SPECIALIZE IN REGULATING **ONCOGENES**-- GENES THAT START AN UNCONTROLLED GROWTH PROCESS THAT CAN LEAD TO CANCER.

LENTIVIRAL VECTORS ARE DEVELOPED STRAIGHT FROM THE HIV AND CAN DELIVER SIGNIFICANT QUANTITIES OF GENETIC INFORMATION, EVEN SPREADING THEIR ALTERED DNA TO NEIGHBORING CELLS THEY HAVE NOT BODILY INVADED. THEY ARE USED TO PROTECT THE IMMUNE SYSTEM AND ARE VALUABLE TOOLS FOR FIGHTING AIDS.

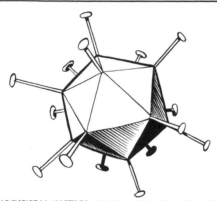

ADENOVIRAL VECTORS, ON THE OTHER HAND, ARE **NOT** RETROVIRUSES-- THEY DO **NOT** INSERT THEIR DNA INTO THEIR HOST'S DNA. THEY EXIST INDEPENDENTLY WITHIN THE CELL, REPRODUCING REGARDLESS OF CELL DIVISION. THEY CAN RELIABLY DELIVER GENES JUST WHERE THEY ARE NEEDED.

NONE OF THESE GENE-THERAPY OPTIONS IS CLOSE TO PERFECT YET. **ONCORETROVIRAL VECTORS,** WHICH PERFORM WELL ONLY IF CELLS ARE DIVIDING, CAN POTENTIALLY CAUSE CANCER THEMSELVES.

LENTIVIRAL VECTORS CAN CAUSE UNWANTED LOSS-OF-FUNCTION MUTATIONS AND CAN REGAIN THEIR DELETED AND DANGEROUS HIV GENES IF THEY COME IN CONTACT WITH UNALTERED HIV VIRUSES.

ADENOVIRAL VECTORS CAN PRODUCE A STRONG REACTION IN A HOST'S IMMUNE SYSTEM AND ARE NOT DEPENDABLY REPLICATED AND PASSED TO DAUGHTER CELLS.

NO ONE'S **PERFECT!**

CURRENT GENETIC THERAPIES CAN TARGET **SPECIFIC CELLS,** AND THERE HAVE EVEN BEEN SUCCESSES ADDRESSING AFFECTED **NUCLEOTIDE SEQUENCES,** ALLOWING THERAPEUTIC GENES TO BE PLACED **ONLY** WHERE THEY ARE REQUIRED. BUT THERE IS A LONG WAY TO GO BEFORE VIRAL GENE THERAPY'S PROMISE IS FULLY REALIZED.

GET ME TO WHERE I CAN DO THE MOST GOOD AND I'LL DO THINGS **RIGHT!**

TO THIS END, **DNA LIBRARIES,** WHICH WILL IDENTIFY AND CATALOG ALL ACTIVE GENES IN A SPECIFIC CELL, ARE BEING DEVELOPED TO MAKE GENE SEARCHES EASIER...

...SINCE THE ALTERNATIVE REQUIRES SEARCHING THE **ENTIRE** HUMAN GENOME FOR EACH SPECIFIC GENE.

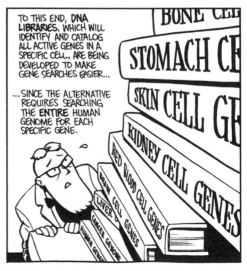

BONE CE
STOMACH CE
SKIN CELL GE
KIDNEY CELL GENES
RED BLOOD CELL GENES
BRAIN CELL GENES
LIVER CELL GENES
CANCER GENOME
HUMAN GENOME

A DNA LIBRARY WILL GUIDE A RESEARCHER **DIRECTLY** TO THE GENES KNOWN TO EXPRESS IN ONLY THE CELLS CONCERNED. THOSE AFFLICTED BY--

OKAY, Bloort-- let's hold up again...

...Because, if I'm understanding this gene-therapy business **CORRECTLY**, it might have some real bearing on our own genetic problems.

Let me try to recap...

...Humans have learned to **IDENTIFY** the source of many disorders and diseases within their genes, and then to **LOCATE** the responsible --or contributing-- genes precisely within their chromosomes.

Knowing where to find specific genes, they've learned to **MANIPULATE** them, first **SPLICING** together bits of DNA into new sequences...

...and **THEN** splicing their own genes within the DNA of **MICROBES** that have the ability to **INSERT** themselves within human cells.

Humans use these microbial **VECTORS** to carry therapeutic human genes into afflicted cells, where the inserted DNA can produce **PROTEINS** to combat the disease!

It's amazing! It's **GENIUS!**

Shouldn't **WE** be able to create a similar therapy to combat the mutating disease that's attacking **US**?!

THAT, I THINK, IS A POSSIBILITY WE **CERTAINLY** MUST EXPLORE, O MOST PERCEPTIVE POTENTATE.

BUT-- THERE IS EVEN **MORE** TO CONSIDER...

CLOSELY RELATED TO RECOMBINANT DNA TECHNOLOGY IS **TRANS-GENIC TECHNOLOGY**, WHICH IS USED TO PRODUCE GMOs--

--GENETICALLY MODIFIED ORGANISMS!

TRANSGENIC TECHNOLOGY INVOLVES LIFTING THE GENES FROM **ONE** ORGANISM AND PLACING THEM WITHIN THE DNA OF **ANOTHER** ORGANISM.

THIS CAN EVEN INCLUDE PLACING **ANIMAL** GENES IN **PLANTS**!

HA! I HEAR YA'.

NOT SURPRISINGLY, THE PUBLIC PERCEPTION OF GMOs IS THAT OF **MONSTERS** RUNNING WILD. WE'LL SEE THAT WHILE THERE MAY BE A SMIDGEN OF REALITY IN THOSE FEARS...

BOO.

... THE **GOALS** ARE MORE IN LINE WITH **IMPROVED MEDICINAL** AND **FOOD RESOURCES**, AND, OF COURSE, **FINANCIAL PROFIT**.

AND **REMEMBER**: HUMANS HAVE BEEN **MODIFYING** PLANTS AND ANIMALS--AND FUNGI--FOR AS LONG AS THEY'VE BEEN AGRICULTURISTS. **CROSSBREEDING** AND EVEN **INDUCED MUTATIONS** ARE RESPONSIBLE FOR CREATING ALL USEFUL DOMESTICATES FROM THEIR WILD ANCESTRY.

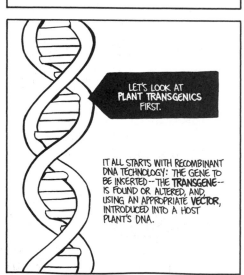

LET'S LOOK AT **PLANT TRANSGENICS** FIRST.

IT ALL STARTS WITH RECOMBINANT DNA TECHNOLOGY: THE GENE TO BE INSERTED-- THE **TRANSGENE**-- IS FOUND OR ALTERED, AND, USING AN APPROPRIATE **VECTOR**, INTRODUCED INTO A HOST PLANT'S DNA.

AGROBACTERIUM, WHICH ENTERS PLANT CELLS TO CAUSE TUMOR-LIKE GROWTHS CALLED **GALLS**, IS A COMMON, USE-FUL VECTOR. LIKE VIRAL OR BACTERIAL VECTORS ALREADY DISCUSSED, **AGRO** HAS ITS UNDESIRABLE GENES CUT FROM ITS DNA PLASMID AND REPLACED WITH THE TRANSGENE.

AGROBACTERIUM   TRANSGENE DNA   { ORIGINAL SOURCES

{ RECOMBINANT AGROBACTERIUM

{ REPLICATED AGROBACTERIA

HOST PLANT CELLS ARE INJECTED WITH THE **ALTERED AGRO**. A PERCENTAGE OF THE INFECTED CELLS GROW INTO MATURE PLANTS, WHICH CARRY THE TRANSGENE SUCCESSFULLY THROUGHOUT THE ENTIRE ORGANISM...

I FEEL **TINGLY** ALL OVER!

...AND EVEN PASS IT ON TO THEIR OFFSPRING.

AGRICULTURAL PRODUCERS ARE EXCITED BY THE POSSIBILITIES OF **HERBICIDE-** AND **PESTICIDE-RESISTANT** CROPS CREATED VIA TRANSGENICS. AN HERBICIDE-RESISTANT SWEET POTATO IS ONE RECENT SUCCESS.

JUST SO LONG AS IT ISN'T **PEST-RESISTANT!**

THE IDEA BEING THAT THESE GENETICALLY "**IMPROVED**" CROPS COULD REMAIN UNAFFECTED BY EVER MORE POTENT WEED AND PEST CONTROLS.

SO-- COUGH-- WE CAN BE SMOTHERED WITH EVEN **MORE** DEADLY CHEMICALS?

KAFF!

EVEN MORE ASTONISHING-- AND **PROBLEMATIC**-- IS THE MEDICINAL POTENTIAL BEING EXPLORED.

HUMAN GENES PRODUCING **TUMOR-FIGHTING ANTIBODY PROTEINS** HAVE BEEN SUCCESSFULLY TRANSFERRED INTO A CROP OF **CORN**...

THE THINGS WE DOMESTICATES DO FOR YOU...

...AND **TOMATOES** IN A RUSSIAN LABORATORY. THEY NOW CARRY FRAGMENTS OF **HIV DNA**, PRODUCING A **VACCINE** THAT, WHEN EATEN, PROMPTS HUMAN GENES TO EXPRESS EFFECTIVE HIV ANTIBODIES!

AND WE MAKE A FINE SALSA.

UNFORTUNATELY, THESE "**PLANTIBODIES**" ARE **ALWAYS** SET TO "**ON**"-- THEY **NEVER** STOP EXPRESSING.

THIS, COUPLED WITH THE ABILITY OF PLANTS TO TRANSFER THEIR GENES **BETWEEN SPECIES**, ALLOWS FOR THE UNCOMFORTABLE POSSIBILITY THAT ENGINEERED TRANSGENES MAY ESCAPE FROM DESIGNATED CROPS AND INFECT LARGER, **UNCONTROLLED** ENVIRONMENTS.

I FEEL **TINGLY** ALL OVER!

THE RESULTS OF UNCONTROLLED TRANSGENE TRANS-FERS ARE UNKNOWN, BUT WOULD MOST LIKELY **NOT** BE DESIRABLE.

ANIMALS, HOWEVER, DO **NOT** NATURALLY TRANSFER GENES BETWEEN SPECIES, SO THAT DANGER DOES NOT EXIST IN **ANIMAL TRANSGENCIES.**

WELL, GOOD FOR ME.

MALE PRONUCLEUS

FEMALE PRONUCLEUS

ZONA PELLUCIDA LAYER

HOLDING NEEDLE

INJECTION NEEDLE

EMBRYO MEMBRANE

BUT, WHILE **ANY PLANT CELL** CAN BE DEVELOPED INTO A MATURE PLANT, **ANIMAL CELLS** MUST HAVE TRANSGENES INSERTED AT AN EARLY STAGE OF DEVELOPMENT, WHILE THEY ARE STILL **TOTIPOTENT.**

THE GREATEST RATE OF SUCCESS IN ANIMAL TRANSGENE INTEGRATION OCCURS WHEN INSERTION IS MADE INTO A **FERTILIZED HOST EGG** DURING THE BRIEF PERIOD BEFORE **PATERNAL** AND **MATERNAL DNA** FUSE.

MANY COPIES OF A TRANSGENE MAY BE INJECTED INTO THE PATERNAL DNA BEFORE -- WITH **LUCK** -- ONE IS SUCCESSFULLY **INTEGRATED.**

MICE WERE THE FIRST ANIMALS IN WHICH A TRANS-GENE -- ONE FOR THE PRODUCTION OF HUMAN GROWTH HORMONE -- WAS SUCCESSFULLY INSERTED.

GIVE ME SOME **MOUSE GROWTH HORMONE** AND THEN WE'LL SEE **WHO** GETS **WHAT** INSERTED INTO THEM!

THE TRANSGENE DOES NOT SPREAD THROUGHOUT AN ANIMAL AS EFFICIENTLY AS IT DOES THROUGH A TOTIPOTENT PLANT. RATHER, IT IS **SLOWLY** AND **ARBITRARILY** REPLICATED IN CHROMOSOMES. FIRST-GENERATION ANIMALS SHOW **RANDOM, INCOMPLETE** TRANSGENE INTEGRATION AND ARE CALLED **CHIMERAS.**

CHIMERAS ARE **CROSSBRED** IN THE HOPE THAT EVENTUALLY A COMPLETELY TRANSGENIC ANIMAL WILL BE PRODUCED.

SOME ANIMALS, BY THE WAY, ACCEPT TRANSGENES BETTER THAN OTHERS.

LUCKY FOR YOU WE'RE ALSO YOUR FAVORITE LAB ANIMALS.

SO **WHY** ENGINEER TRANSGENES INTO ANIMALS?

TO BETTER STUDY **HUMAN DISORDERS** WITHOUT EXPERI-MENTING ON HUMANS THEMSELVES, FOR ONE THING...

WITH **85%** OF OUR GENES EQUIVA-LENT, WE'RE TOO CLOSELY RELATED FOR **MY** COMFORT!

...AND, SIMILAR TO **PLANT** TRANSGENICS, THE HOPE IS THAT HEALTHIER, CHEAPER, AND MORE CONSISTENT **FOODS** AND **MEDICINES** CAN BE PRODUCED.

A TRANSGENE CODING FOR A SPECIFIC **ENZYME CONVERTER**, INSERTED INTO **PIGS**, CONVERTS DIET-HARMFUL **OMEGA-6 FATTY ACIDS** INTO HEALTHY **OMEGA-3 FATTY ACIDS!**

HEY, WE'RE TALKING **HEALTHY BACON** HERE!

COWS NOW PRODUCE **HUMAN PROTEIN C** IN THEIR MILK. GOOD NEWS FOR HEMOPHILIACS AND FOR **STROKE** AND **HEART ATTACK** VICTIMS!

TRANSGENICS ALSO SHOWS PROMISE AS A MEANS OF CONTROLLING **DISEASE-BEARING** VECTORS.

TRANSGENES INSERTED INTO **MALARIAL MOSQUITOES** CAN RENDER THE INSECTS **INCAPABLE** OF CARRYING THE PARASITE RESPONSIBLE FOR THAT DREAD DISEASE.

GREAT! WHO WANTS PARASITES?

AND TRANSGENES FROM **THREE DIFFERENT ANIMAL SOURCES** HAVE BEEN INSERTED INTO **YEAST** TO HELP FABRICATE A **DRUG** TO FIGHT MALARIA.

THIS PARTICULAR PROJECT HAS **ALREADY PROVEN A SUCCESS!**

HMMM... MAYBE ANTI-MALARIAL **RYE** SOMEDAY...

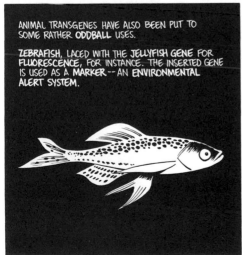

ANIMAL TRANSGENES HAVE ALSO BEEN PUT TO SOME RATHER **ODDBALL** USES.

ZEBRAFISH, LACED WITH THE **JELLYFISH GENE** FOR **FLUORESCENCE**, FOR INSTANCE. THE INSERTED GENE IS USED AS A **MARKER** -- AN **ENVIRONMENTAL ALERT SYSTEM.**

THE TRANSGENE IS MANIPULATED TO EXPRESS -- TO **FLUORESCE** -- ONLY WHEN **POLLUTANTS** IN ITS ZEBRAFISH HOST'S WATERS REACH A SPECIFIED **UNHEALTHY LEVEL.**

EVERYONE OUT OF THE POOL!

MANY HUMANS ARE UNDERSTANDABLY NERVOUS ABOUT THE POTENTIAL DANGERS GENETICALLY MODIFIED ORGANISMS MAY PRESENT.

BUT THAT NERVOUSNESS IS **NOTHING** COMPARED TO THE SWAMP OF **FEAR** AND **ETHICAL CONTROVERSY** SWIRLING AROUND **ANOTHER** ASPECT OF GENETIC ENGINEERING...

... CLONING TECHNOLOGY!

CLONES!!

A **CLONE** IS AN **EXACT DUPLICATE** OF AN ORGANISM AND, AS WITH THE POPULAR NOTIONS OF **MUTANTS**, CLONES HAVE BECOME A BYWORD FOR **SCIENCE RUN AMOK**...

...THIS DESPITE THE FACT THAT CLONES HAVE BEEN A **BIG PART** OF LIFE ON EARTH SINCE ITS BEGINNINGS.

YOUR OLD FRIENDS, BACTERIA, FOR INSTANCE, ARE A SIZABLE **MAJORITY** OF CLONES!

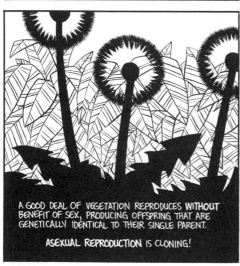

A GOOD DEAL OF VEGETATION REPRODUCES **WITHOUT** BENEFIT OF SEX, PRODUCING OFFSPRING THAT ARE GENETICALLY IDENTICAL TO THEIR SINGLE PARENT.

ASEXUAL REPRODUCTION IS CLONING!

SOME **ANIMALS** CAN DO IT THIS WAY, TOO. IN THIS KINGDOM, ASEXUAL REPRODUCTION IS COMMONLY REFERRED TO AS **PARTHENOGENESIS**.

'SNUTHIN' NEW!

LITTLE FIRE ANT

HYDRA

WITHIN **HUMAN CULTURE**, CLONING TECHNIQUES HAVE BEEN USED SINCE THE EARLIEST DAYS OF AGRICULTURE. FRUIT FARMERS PASS CONSISTENT, DESIRABLE TRAITS FROM ONE GENERATION OF CROP TO THE NEXT BY **GRAFTING** A CUTTING FROM THE PREVIOUS GENERATION ONTO NEW ROOT STOCK.

THE TRAITS OF THE OLD STOCK ARE **TRANSFERRED** TO THE NEW AS THE TWO FUSE AND THE DNA OF THE OLD INFORMS THE NEW.

GRAFTING IS CLONING!

JOHN CHAPMAN WAS A PIONEER OF THE AMERICAN WEST WHO GAVE THE **NEW WORLD** ITS OWN HEALTHY DIVERSITY OF APPLE VARIETIES. POPULARLY KNOWN AS **JOHNNY APPLESEED**, HE HAS BEEN **MYTHOLOGIZED** AND PRACTICALLY **CANONIZED** AS AN AMERICAN FOLK HERO. BUT THE ROUGH-HEWN FACTS OF WHAT HE ACCOMPLISHED ARE OFTEN WHITEWASHED TO THE POINT WHERE AN IMPORTANT LESSON IS LOST.

HOW THE WEST WAS SEEDED
(instead of cloned)

CHAPMAN HELPED DOMESTICATE THE AMERICAN FRONTIER IN THE EARLY YEARS OF THE 19th CENTURY BY PLANTING **APPLE SEEDS** WELL IN ADVANCE OF THE WESTWARD-SPREADING POPULACE. HE WAS A SORT OF **LAND SPECULATOR** -- HE WOULD SCOPE OUT NEW HORIZONS, DETERMINE WHERE SETTLEMENTS COULD BE MADE, AND THERE PLANT NEW ORCHARDS. BY THE TIME THE SETTLERS ARRIVED, AN ORCHARD WOULD BE WAITING. HE DID VERY WELL AT HIS BUSINESS (ALTHOUGH HIS ECCENTRIC, VAGABOND LIFESTYLE BELIED HIS FINANCIAL WORTH).

APPLE TREES ARE, BY NATURE, THE **EXTREME OPPOSITE** OF A TRUE-BREEDING PLANT. WHEN THEY ARE PLANTED FROM SEED, ONLY VERY RARELY WILL THE FRUIT OF THE OFFSPRING RESEMBLE THAT OF THE PARENT. APPLE TREES, NATIVES OF CENTRAL ASIA, ARE ECOLOGICALLY SUCCESSFUL PLANTS **BECAUSE** OF THIS STRATEGY OF EXTREME DIVERSITY.

HUMANS, HOWEVER, DESIRE A HIGH DEGREE OF CONSISTENCY IN THEIR CROPS. THE CHINESE DISCOVERED **GRAFTING 4,000 YEARS AGO**, AND IT WAS THIS CLONING TECHNIQUE THAT ALLOWED AGRICULTURISTS TO CREATE **CONSISTENT** VARIETIES OF APPLES. THE PRACTICE OF GRAFTING TO MAINTAIN STANDARDIZED FRUITS WAS PASSED ON TO EUROPEANS, WHO, IN TURN, BROUGHT BOTH THE APPLE AND THE PRACTICE TO THE WESTERN HEMISPHERE.

GRAFTED OLD-WORLD APPLE TREES DID **NOT** TEND TO FARE WELL UNDER AMERICAN ENVIRONMENTAL CONDITIONS. IT TOOK MEN LIKE CHAPMAN, WHO WERE WILLING TO GO BACK TO SEED, TO FIRMLY ESTABLISH THE EXOTIC APPLE AS A ROBUST, NATURALIZED TREE IN AMERICA. CHAPMAN WAS A DEEPLY RELIGIOUS MAN (ALTHOUGH THE EXACT NATURE OF HIS SPIRITUAL BELIEFS IS UP FOR DEBATE) WHO VIEWED GRAFTING AS **UNNATURAL** AND WOULD HAVE NOTHING TO DO WITH IT.

OF COURSE, THE APPLES FROM SEED THAT CHAPMAN PLANTED FOLLOWED THEIR TRUE NATURE AND EXPRESSED THEIR INCREDIBLE GENETIC DIVERSITY, PRODUCING A VAST MAJORITY OF BITTER OR SOUR FRUITS NOT PLEASANT TO HUMAN TASTE. SO, WHAT **GOOD** WERE ALL THOSE ORCHARDS THAT CHAPMAN PLANTED?

NOT MUCH -- IF YOU ARE CONSIDERING APPLES AS HUMANS DO TODAY. BUT IN CHAPMAN'S TIME APPLES WERE PRIMARILY USED TO MAKE AN ALCOHOLIC BEVERAGE -- **CIDER**. IT WAS **MORE** THAN A MERE INTOXICANT, TOO -- WITH WATER QUALITY BEING AT BEST AN IFFY PROPOSITION, CIDER WOULD BE A MUCH PREFERRED ALTERNATIVE TO CHOLERA.

APPLES WERE LARGELY REGARDED AS CIDER STOCK UNTIL THE **1920s**, WHEN THE ANTI-ALCOHOL **TEMPERANCE** MOVEMENT THAT CULMINATED IN **PROHIBITION** DEMONIZED ALL DRINKS ALCOHOLIC. ONLY THEN DID THE APPLE-GROWING INDUSTRY REINVENT ITSELF AS PURVEYORS OF A TASTY, NUTRITIOUS FRUIT TO BE EATEN RAW, COINING THE PHRASE "AN APPLE A DAY KEEPS THE DOCTOR AWAY."

BUT NOW IT WAS TO THE INDUSTRY'S ADVANTAGE TO PRODUCE CONSISTENT FRUIT THAT TASTED **RELIABLY SWEET** AND COULD BE **EFFICIENTLY SHIPPED**. GRAFTING BECAME STANDARD PROCEDURE FOR PROPAGATING AMERICAN APPLES, AND DIVERSITY WAS REDUCED TO THE HANDFUL OF **CLONES** SEEN IN SUPERMARKETS TODAY, ALL RELYING ON EVER-INCREASING CHEMICAL ASSISTANCE TO BATTLE **CATASTROPHIC DEVASTATION** FROM THE RAPIDLY EVOLVING **INSECTS, FUNGI,** AND **MICROBES** THAT HAVE GAINED THE EVOLUTIONARY UPPER HAND.

IF THERE IS A GREAT APPLE ORCHARD IN THE SKY, JOHN CHAPMAN MUST BE LOOKING DOWN, NOT AT **ALL** HAPPY.

A COMMERCIAL VARIETY OF APPLE -- A **MACINTOSH**, FOR INSTANCE -- THAT A HUMAN EATS TODAY IS **GENETICALLY IDENTICAL** TO A MACINTOSH EATEN 30 YEARS AGO!

STILL TASTES FRESH TO ME.

BUT THAT'S MERE **RECENT** HISTORY! THERE ARE GRAPE CULTIVARS IN EUROPE THAT HAVE BEEN PROPAGATED BY CLONING FOR OVER 2,000 YEARS!

AND YOU **BET** WE'RE FEELING OUR AGE...

THE HUMAN SPECIES **ITSELF** IS RIFE WITH CLONES -- BUT HUMANS REFER TO THEM AS **IDENTICAL TWINS.**

WE HAVE ALWAYS LIVED AMONG YOU, BIDING OUR TIME...

WHOA! Wait a minute -- this clone business...

...This means we Squinch are **CLONES**, right?

WE reproduce without benefit of sex...

...thank goodness...

THAT'S RIGHT -- IT'S PARTHENOGENESIS FOR US.

WE ARE CLONES.

AND A VERY **HYGIENICALLY SOUND** STRATEGY, IN MY OPINION.

So what could humans **POSSIBLY** find frightening about **CLONES**??

I mean, **LOOK** at us! We're **ADORABLE!**

WELL, LET ME BE PERFECTLY CLEAR -- IT WAS, AS USUAL, ADVANCES IN SCIENTIFIC KNOWLEDGE AND TECHNOLOGY THAT CREATED THE UPROAR.

IN 1952, HUMANS FIRST ARTIFICIALLY CLONED AN ANIMAL -- A FROG TADPOLE -- FROM EMBRYONIC CELLS.

BUT IT WASN'T UNTIL 1996 AND THE FIRST SUCCESSFUL CLONING OF A MUCH MORE COMPLEX ANIMAL -- A MAMMAL -- THAT THE FUROR BEGAN IN EARNEST.

MAMMALS ARE SO CUTE! THEY'RE JUST LIKE US!

DOLLY THE SHEEP WAS THE FIRST MAMMAL CLONED FROM AN ADULT CELL THAT LIVED PAST BIRTH.

HER COST WAS EXTRAORDINARILY HIGH. SHE WAS THE ONLY SUCCESS IN A PROJECT ATTEMPTING TO PRODUCE CLONES IN 277 SHEEP EGGS.

DOLLY DIED, PREMATURELY, AT AGE SIX -- A DISAPPOINTINGLY BRIEF LIFE FOR A SHEEP. WHETHER THE CAUSE OF DEATH WAS RELATED TO HER CLONED NATURE OR NOT IS STILL A MATTER OF DEBATE.

SINCE DOLLY, THERE HAVE BEEN OTHER SUCCESSES -- AND FAILURES. BUT MORE FAILURES THAN SUCCESSES.

THE EXCITEMENT OVER CLONING HAS, IN MANY REGARDS, BEEN SUPPLANTED BY ADVANCES IN RECOMBINANT DNA TECHNOLOGY.

TRUTH BE TOLD, RECOMBINANT DNA TECHNOLOGY IS REALLY JUST CLONING ON A MOLECULAR LEVEL!

REPRODUCTIVE CLONING -- REPRODUCING WHOLE ANIMALS -- HAS REMAINED TOO EXPENSIVE AND PROBLEMATIC TO MAKE GOOD MEDICAL OR COMMERCIAL SENSE...

...ALTHOUGH THERE ARE PARTIES WHO REMAIN HOPEFUL.

WHAT CARNIVORE WOULDN'T WANT A GUARANTEED PERFECT GRADE A SIRLOIN STEAK FOR EVERY MEAL?

MAYBE EVEN WITH OMEGA-3 FATTY ACIDS!

ANOTHER USE FOR CLONING THAT **MAY** HAVE A FUTURE IS THAT OF **THERAPEUTIC TISSUE REPLICATION** -- THE CLONING OF INDIVIDUAL **ORGANS** FOR TRANSPLANT PURPOSES.

IT BEATS WAITING FOR A DONOR.

HUMAN EGG CELL

TISSUE CLONED FROM THE PATIENT'S DNA WOULD **AVOID** THE DANGER OF DONOR-ORGAN **REJECTION** BY THE IMMUNE SYSTEM.

THE SCENARIO WORKS LIKE THIS: DNA FROM AN INDIVIDUAL NEEDING A REPLACEMENT ORGAN IS INSERTED INTO A **HUMAN EGG** STRIPPED OF ITS **NUCLEUS.**

AFTER THE EGG, WITH THE TRANSFERRED DNA, BEGINS TO DIVIDE, **EMBRYONIC STEM CELLS** -- CELLS THAT ARE AT THIS STAGE CAPABLE OF DEVELOPING INTO ANY TYPE OF TISSUE THAT THE BODY NEEDS -- ARE **HARVESTED** AND **GUIDED** TO GENERATE THE SPECIFIC ORGAN TISSUE NEEDED.

SUDDENLY, I FEEL LIKE BECOMING A HEART...

THE NEW ORGAN WOULD BE A **CLONE** OF THE PATIENT'S ORIGINAL.

IF THIS TECHNOLOGY BECOMES FEASIBLE, THE **SAFETY** AND **SUCCESS RATE** FOR TRANSPLANTATION WOULD BE GREATLY **INCREASED,** WHILE THE NEED FOR **ORGAN DONATION** WOULD BE DRASTICALLY **REDUCED.**

BUT...

YEP -- THE SPITTING IMAGE OF MY ORIGINAL EQUIPMENT!

... AS RESEARCH IS NOW CONDUCTED THROUGH A PROCESS THAT **KILLS** THE HUMAN EGG -- A POTENTIAL HUMAN LIFE -- WHEN STEM CELLS ARE HARVESTED, VARIOUS **INTEREST GROUPS** HAVE SOUGHT TO **RESTRICT** OR **OUTLAW** THEIR USE.

THE **ETHICAL** OR **SUPERSTITIOUS** OBJECTIONS THESE INTEREST GROUPS RAISE PLACE LIMITATIONS ON THE CONTINUATION OF STEM CELL RESEARCH...

... ALTHOUGH THE OBJECTIONS MIGHT BE OVERCOME BY RECENT DEVELOPMENTS INDICATING THAT THERE MAY BE A MEANS OF HARVESTING THESE VALUABLE CELLS **WITHOUT** DAMAGING THE EGG.

ONLY TIME WILL TELL HOW THIS DRAMA OF MEDICAL ADVANCEMENT VERSUS MORAL CONCERN ENDS!

WHILE **TISSUE CLONING** STILL SHOWS **POTENTIAL**, HOWEVER, SCIENTIFIC INTEREST IN THE POSSIBILITIES OF **REPRODUCTIVE CLONING** SEEMS TO HAVE WANED. ALTHOUGH NEW SPECIES ARE INDEED CLONED EVERY YEAR...

...THE ENORMOUS **EXPENSE** AND THE **HIGH RATE** OF FAILURE CONTINUE TO BE ROADBLOCKS.

A **QUALIFIED** BRIGHT SPOT: CONSERVATIONISTS HAVE LOOKED TO CLONING AS A LAST-DITCH, WORST-CASE SCENARIO FOR SAVING SPECIES ON THE BRINK OF EXTINCTION.

WELL, IT WORKED FOR **ME**...

YEAH, BUT TO LIVE ON WITHOUT THE ENVIRONMENT IN WHICH I EVOLVED...

...I DON'T KNOW...

MOUFLON   SUMATRAN TIGER

AND **FINALLY**: THE IDEA THAT GENETIC ENGINEERING WILL PRODUCE THE PERFECT **BABY-TO-ORDER**?

THE NOTION THAT CLONING WILL KEEP A **BELOVED PET** WITH US **FOREVER**?

THAT'S **NOT** GOING TO HAPPEN. EVER.

CLONES ARE EXACT DUPLICATES IN **GENOTYPE** ONLY. THERE ARE **FAR** TOO MANY **ENVIRONMENTAL** AND **MUTATIONAL** FACTORS AT WORK IN PRODUCING THE **PHENOTYPE** OF A COMPLEX ORGANISM TO **EVER** PRODUCE AN EXPRESSLY CRAFTED DUPLICATE-ON-DEMAND.

I CAN'T **BELIEVE** HOW MY ORIGINAL BEHAVED! WHAT KIND OF **UPBRINGING** DID HE HAVE?

TELL me about it! None of **MY** offspring are exactly chips off the old block.

BUT WE ARE ALL **GRATEFUL** FOR THAT, I ASSURE YOU.

All joking aside...

...I am well pleased with what you have taught me today, Bloort 183. Your mission to Earth was undeniably worthwhile. You have provided us with some important lessons--

--as alien and hard to digest as they may be.

But, tell me, just how is it that we Squinch have never explored our own biology so thoroughly?

MAYBE, O WAKENING GIANT, BECAUSE WE EVOLVED SO **SLOWLY** WHILE SOMEHOW MANAGING TO AVOID A MASSIVE CRISIS SUCH AS NOW CONFRONTS OUR SPECIES.

BECAUSE WE HAVE BEEN **SO** STABLE, MAYBE THE NOTION OF CHANGE NEVER OCCURRED TO US.

BUT EVEN THOUGH WE ARE ASEXUAL BEINGS WITHOUT THE MECHANICS FOR NIMBLE ADAPTATION, WE **HAVE** EVOLVED...

...SLOWLY AND PONDEROUSLY, THROUGH GRADUAL, PERSISTENT MUTATION.

WE DO GENETICALLY CHANGE.

WHAT WE MUST DO **NOW** IS FIND A WAY OF **HARNESSING** WHAT WE HAVE **WITHIN** OUR BIOLOGY IN ORDER TO BEAT THIS CRISIS!

Well, the **HUMANS** are developing a **VARIETY** of strategies for curing their genetic ills.

BUT THOSE WON'T NECESSARILY WORK FOR **US**...

WE MUST FIRST EXPLORE **OUR** GENETIC HISTORY -- THE GENETIC HISTORY OF **OUR** WORLD -- TO UNDERSTAND WHERE **WE** COME FROM.

TO UNDERSTAND WHY WE ARE NOW FAILING BEFORE THIS...UNFORTUNATE **DISORDER**.

Very well, then. We must learn our genetic past.

Even down to the-- sigh-- Frippurvian flogworm.

So tell us, Bloort -- just how **ARE** the humans uncovering **THEIR** history?

CHAPTER 5

Applying All That Stuff-- Into the Future Through the Past

MODERN HUMAN BEINGS-- *HOMO SAPIENS* IN SCIENTIFIC NOMENCLATURE-- ARE A **SINGLE SPECIES**, BUT ONE WITH PLENTY OF VARIATION.

ALL THIS VARIATION IS LARGELY **PHENOTYPICAL**, HOWEVER. EVERY HUMAN'S PERSONAL GENOME-- WHICH CONTAINS THEIR **GENOTYPE**-- DIFFERS FROM ANOTHER'S BY ONLY A FEW HUNDREDTHS OF A PERCENTAGE POINT.

REMEMBER-- 98% OF THE GENES OF HUMANS AND THEIR CLOSEST LIVING RELATIVES, CHIMPANZEES, ARE PRACTICALLY IDENTICAL!

WHILE HUMANS BECAME ANATOMICALLY MODERN SOMEWHERE AROUND 150,000 TO 200,000 YEARS AGO, WITH THEIR MODERN **BEHAVIOR PATTERNS** APPEARING ONLY ABOUT 50,000 YEARS AGO, THE VARIATIONS WE SEE IN DIFFERENT POPULATIONS SCATTERED ACROSS THE EARTH HAVE APPEARED **MUCH** MORE RECENTLY.

AND DIFFERING **PHYSICAL** CHARACTERISTICS ONCE CONSIDERED TO DEFINE **RACES** (AN OUTMODED CONCEPT) APPEARED AS RECENTLY AS ONLY 5,000 YEARS AGO!

BUT THE VARIATIONS IN THE HUMAN SPECIES **ARE** SIGNIFICANT. THEY REFLECT THE FACT THAT HUMANS MIGRATED FROM A **SINGLE POINT OF ORIGIN** TO ESTABLISH GENETICALLY DIFFERENTIATED POPULATIONS AROUND THE GLOBE.

EACH ISOLATED POPULATION FACED ITS OWN ENVIRONMENTAL AND CULTURAL CHALLENGES AND IN RESPONSE EVOLVED ITS OWN SET OF UNIQUE GENETIC ASSETS -- AND LIABILITIES.

DNA REVEALS THE STILL-DEVELOPING STORY OF THE HUMAN SPECIES!

HUMANS HAVE A PRETTY GOOD IDEA OF THEIR HISTORY GOING BACK 5,000 YEARS, WHICH INCLUDES THE EARLIEST KNOWN WRITTEN RECORDS.

THE MOST ANCIENT EVIDENCE THEY'VE YET DISCOVERED FOR HUMAN SETTLEMENTS IS ABOUT **40,000 YEARS OLD,** BUT THE SPECIFICS GOT INCREASINGLY FOGGY ANY EARLIER THAN THAT...

...UNTIL GENETIC RESEARCHERS BECAME **ANTHROPOLOGICAL DETECTIVES** AND BEGAN PIECING TOGETHER THE PUZZLE OF EARLY HUMAN BIOLOGICAL, SOCIAL, AND CULTURAL EVOLUTION.

THE SEARCH FOR THE ORIGINS OF THE MODERN HUMAN SPECIES HAS BENEFITED GREATLY FROM **TWO GENETIC FACTORS...**

...ONE IS THE **Y CHROMOSOME** -- CARRIED BY MALES ONLY...

...AND THE OTHER IS THE DNA FOUND WITHIN **MITOCHONDRIA,** UNIQUE TO THAT CELLULAR ORGANELLE...

...AND PASSED EXCLUSIVELY FROM **MOTHER** TO **CHILD.**

TRACING HUMAN HISTORY THROUGH THE Y CHROMOSOME IS MADE EASIER BY THE FACT THAT ALL MEN TODAY CARRY A Y CHROMOSOME INHERITED FROM A **SINGLE ANCESTOR!**

WHY? BECAUSE THAT ANCIENT MAN'S LINEAGE OUTLASTED ALL OF HIS APPROXIMATELY 2,500 MALE COMPETITORS!

THROG'S MALE SEED IS GOING NOWHERE-- HE'S GIVEN LIFE ONLY TO DAUGHTERS!

BUT, AS WE'VE SEEN ELSEWHERE, INHERITED CHROMOSOMES ARE **NOT** IDENTICAL COPIES. MUTATIONS ARE CONSTANTLY OCCURRING WITHIN GENES.

ONCE A MUTATION APPEARS ON A Y CHROMOSOME, IT IS HANDED DOWN THROUGH SUCCESSIVE GENERATIONS, FATHER TO SON.

THIS Y CHROMOSOME IS YOUR **INHERITANCE,** SON. I RECEIVED IT FROM **MY** FATHER, BUT THERE HAVE BEEN SOME IMPROVEMENTS MADE...

AS MUTATIONS OCCUR ON THE Y CHROMOSOME, **FORKS** ARE CREATED IN THE HUMAN FAMILY TREE BETWEEN THOSE CARRYING A PARTICULAR MUTATION AND THOSE WITHOUT IT. THESE FORKS FROM THE ROOT ANCESTOR CREATE NEW **LINEAGES**.

M170 MUTANTS

THE APPEARANCE OF A PARTICULAR MUTATION ON A MAN'S Y CHROMOSOME CAN TELL A **POPULATION GENETICIST** TO WHAT **LINEAGE** THAT MAN BELONGS.

SO, THE SPECIFIC SET OF MUTATIONS WITHIN A Y CHROMOSOME DETER-MINES THE OWNER'S **LINEAGE**, AND THE **TIME** THAT LINEAGE FORMED...

...AND HAS HELPED SCIENTISTS **GEOGRAPHICALLY TRACE** THE SPREAD OF HUMANKIND FROM ITS ANCESTRAL HOME AND ACROSS THE PLANET.

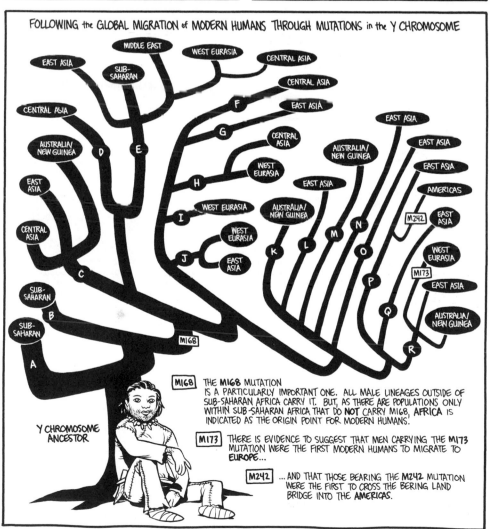

FOLLOWING the GLOBAL MIGRATION of MODERN HUMANS THROUGH MUTATIONS in the Y CHROMOSOME

M168 THE **M168** MUTATION IS A PARTICULARLY IMPORTANT ONE. ALL MALE LINEAGES OUTSIDE OF SUB-SAHARAN AFRICA CARRY IT. BUT, AS THERE ARE POPULATIONS ONLY WITHIN SUB-SAHARAN AFRICA THAT DO **NOT** CARRY M168, **AFRICA** IS INDICATED AS THE ORIGIN POINT FOR MODERN HUMANS.

M173 THERE IS EVIDENCE TO SUGGEST THAT MEN CARRYING THE **M173** MUTATION WERE THE FIRST MODERN HUMANS TO MIGRATE TO **EUROPE**...

M242 ...AND THAT THOSE BEARING THE **M242** MUTATION WERE THE FIRST TO CROSS THE BERING LAND BRIDGE INTO THE **AMERICAS**.

Y CHROMOSOME ANCESTOR

SCIENTISTS ARE ABLE TO CALCULATE THE **RATE** AT WHICH MUTATIONS ACCUMULATE. THE M168 MUTATION'S APPEARANCE-- SOON BEFORE MODERN HUMANS BEGAN THEIR SUCCESSFUL TREK OUT OF AFRICA--

M168

-- CAN BE DATED TO ABOUT 44,000 YEARS AGO...

... WHILE THE GENETIC **"ADAM"** WHO CONTRIBUTED THE MODERN HUMAN Y CHROMOSOME IS ESTIMATED TO HAVE LIVED ABOUT **59,000 YEARS AGO.**

THESE DNA-DERIVED DATES ARE ONLY APPROXIMATE, BUT THEY FIT NICELY WITH PALEONTO-LOGICAL EVIDENCE THAT ALSO PLACES GREAT ADVANCES IN MODERN HUMAN BEHAVIOR AT AROUND **50,000 YEARS AGO.**

WHAT'S A FEW THOUSAND YEARS? I'M **STILL** YOUR FATHER!

MITOCHONDRIAL DNA SEEMS TO CONFIRM THIS HISTORY OF HUMAN DISPERSION AS WELL.

LIKE THE Y CHROMOSOME LINEAGE, ALL MITOCHONDRIAL DNA COMES FROM A **SINGLE** ROOT SOURCE...

... A MITOCHONDRIAL **"EVE"** WHOSE LINEAGE OUTLASTED ALL HER COMPETING MATRIARCHAL LINEAGES.

LIKE CHROMOSOMAL DNA, MITOCHONDRIAL DNA COLLECTS **MUTATIONS,** ALLOWING SCIENTISTS TO CALCULATE DATES...

... ALTHOUGH WITH SOME-WHAT LESS **ACCURACY** THAN WITH THE Y CHROMOSOME, AS MITOCHONDRIAL DNA MUTATES MORE **QUICKLY.**

EVEN SO, THE TIME FRAME FOR THE MOTHER OF MODERN HUMAN MITOCHONDRIAL DNA CAN BE APPROXIMATED AT ABOUT **150,000 YEARS AGO,** SIGNIFICANTLY OLDER THAN THE Y CHROMOSOME FATHER.

I AM THE MOTHER OF ALL MOTHERS! AND ALL FATHERS!

BASED ON THE NUMBER OF ALLELIC VARIATIONS SEEN IN BOTH CONTEMPORARY Y CHROMOSOMES AND MITOCHONDRIAL DNA AND THE KNOWN RATE OF MUTATION ACCUMULATION, THE SIZE OF THE ANCESTRAL MODERN HUMAN POPULATION IS ESTIMATED TO BE AROUND ONLY **5,000 INDIVIDUALS!**

2,500 MALES...

...AND 2,500 FEMALES. AND ONLY **ONE** MALE Y CHROMOSOME LINEAGE AND **ONE** FEMALE MITOCHONDRIAL LINEAGE WILL SURVIVE TO YOUR TIME!

FOLLOWING the GLOBAL MIGRATION of MODERN HUMANS THROUGH MUTATIONS in MITOCHONDRIAL DNA

AS WITH MALE LINEAGES TRACED THROUGH THE Y CHROMOSOME, FORKS FROM THE ANCESTRAL FEMALE ROOT ARE CREATED BY **MUTATION**. CERTAIN EARLY BRANCHES ON THE MITOCHONDRIAL DNA TREE ARE FOUND ONLY IN **SUB-SAHARAN AFRICA**...

... ADDING CONFIRMING EVIDENCE TO THE THEORY THAT **AFRICA** WAS THE ORIGINAL HOME OF MODERN HUMANS. THE **M** AND **N** BRANCHES REPRESENT THE LINEAGES THAT MIGRATED **OUT OF AFRICA** TO POPULATE THE REST OF THE PLANET.

THE TINY POPULATION OF 5,000 EARLY MODERN HUMANS WAS PROBABLY THE REMAINS OF A MUCH LARGER POPULATION DECIMATED BY CATASTROPHIC ENVIRONMENTAL EVENTS.

WHATEVER THOSE UNKNOWN CATASTROPHES MIGHT HAVE BEEN, CONDITIONS AT ABOUT **50,000 YEARS AGO** MUST HAVE EXISTED THAT FORCED AT LEAST ONE BAND OF AS FEW AS **150 INDIVIDUALS** ON THE EXTREMELY DANGEROUS TREK OUT OF THE ANCESTRAL HOMELAND.

PLEASE KEEP IN MIND THAT THE HUMAN GENOME IS, OF COURSE, **CONSTANTLY** EVOLVING.

THERE IS **NO** ONE POINT IN TIME OR ANY **SPECIFIC** MUTATION THAT LITERALLY SEPARATES MODERN HUMANS FROM THEIR ANCESTORS.

THE 50,000-YEARS-AGO DATE USED TO DESCRIBE THE APPEARANCE OF MODERN HUMANS IS **ARBITRARY**, RELATED NOT SO MUCH TO ANY SPECIFIC BIOLOGICAL DEVELOPMENT BUT MORE TO RECOGNITION OF THE SIGNIFICANCE OF THE EVENTS OF THE TIME...

...MOST IMPORTANT, THE SUCCESSFUL **TREK** OUT OF AFRICA.

THERE IS ARCHAEOLOGICAL EVIDENCE OF **EARLIER** ATTEMPTS THAT **FAILED**, POSSIBLY THE RESULT OF DIRECT CONFLICT WITH HOSTILE FORCES OCCUPYING LANDS THAT **BLOCKED** THE ROUTE FROM THE HOMELAND.

NO COME BACK!

MOST LIKELY, THE HOSTILES WERE MODERN HUMANS' EXTINCT RELATIVE, *HOMO NEANDERTHALENSIS* -- **NEANDERTHAL MAN.**

WHY DID MODERN HUMANS **FINALLY** SUCCEED IN BREAKING OUT OF AFRICA?

GENETIC RESEARCH HAS DISCOVERED CHANGES IN THE HUMAN BRAIN THAT MOST LIKELY LED TO THE DEVELOPMENT OF IMPORTANT ADVANTAGES AT ABOUT THIS TIME: SOPHISTICATED **LINGUISTICS** AND **SOCIAL STRUCTURES.**

...AND SO, MY DEAR FELLOWS, I SAY WE MUST ALL HANG **TOGETHER** -- OR SURELY WE WILL ALL HANG **SEPARATELY**...

Panel 3:

TO FIND THE ORIGINS OF THESE GENETIC ADVANCES WE NEED TO LOOK **FURTHER** BACK, TO THE TIME WHEN THE LINEAGES OF **PROTO-HUMANS** AND THEIR CLOSEST RELATIVES, **CHIMPANZEES**...

CHIMPANZEE

...SEPARATED FROM THEIR ANCIENT COMMON ANCESTOR.

ANCESTRAL PRIMATE

PROTO-HUMAN

Panel 4:

HUMANS AND CHIMPS HAVE BETTER THAN 98% OF THEIR GENES ANALOGOUS OR HOMOLOGOUS. BY COMPARING THE NUMBER OF **DIFFERENCES** IN CORRESPONDING STRETCHES OF HUMAN AND CHIMP DNA, GENETICISTS ARE ABLE TO DATE THE SPLIT BETWEEN THE TWO SPECIES AT APPROXIMATELY **5 MILLION** YEARS AGO.

GOODBYE! LET'S NOT LET OUR DIFFERENCES MAKE US STRANGERS!

Panel 5:

SEVERE CLIMATIC SHIFTS AT THAT TIME WERE CAUSING **MASSIVE ENVIRONMENTAL CHANGES** IN AFRICA. INDIVIDUALS WITH MUTATIONAL VARIATIONS BEST SUITED TO **COPING** WITH THE NEW, MORE **ARID** CONDITIONS WERE MOST LIKELY TO SURVIVE -- AND SO PASS ON THEIR GENES TO FUTURE GENERATIONS.

OUT OF THE FOREST AND INTO THE GRASS! AN EXCELLENT EXAMPLE OF PRACTICAL EVOLUTION!

Panel 6:

ONE OF THE MOST IMPORTANT MUTATIONS -- A REAL ASSET AS HUMANS BRANCHED OUT TO CONQUER A NEW ENVIRONMENTAL NICHE -- INVOLVED THE DEVELOPMENT OF THE **BRAIN.**

GENETICISTS, AGAIN COMPARING CHIMP AND HUMAN GENOMES, HAVE FOUND SPECIFIC STRETCHES OF DNA ENCOURAGING **RAPID GENE MUTATION** ONLY IN THE **HUMAN!**

I CAN'T BELIEVE WE USED TO HANG OUT WITH THOSE GUYS.

THESE HUMAN-ACCELERATED REGIONS, OR **HARS**...

...OFFER CLUES TO HUMAN BRAIN EVOLUTION. MANY OF THE HARS ARE LOCATED NEAR GENES SPECIFICALLY INVOLVED WITH BRAIN DEVELOPMENT.

HAR 1 LIES WITHIN A GENE THAT CODES FOR AN RNA MOLECULE THAT MAY SUPERVISE BRAIN CONSTRUCTION IN THE WOMB.

PRIORITIZE ACCELERATED DEVELOPMENT

HAR 1-- SIGNIFICANTLY **NOT** PRESENT IN THE CHIMP GENOME-- MAY REPRESENT AN IMPORTANT MUTATION AT LEAST PARTIALLY RESPONSIBLE FOR HUMANKIND'S UNIQUE ADVANCEMENT INTO INTELLECTUAL SOPHISTICATION.

IMPROVED INTELLECTUAL CAPACITY QUITE POSSIBLY LED TO MUCH MORE **COMPLEX SOCIAL STRUCTURE**, INCLUDING MALE-FEMALE PAIR BONDING, AND LONG-TERM FAMILY RELATIONSHIPS THAT ENGENDERED A GREATER INTEREST IN GROUP WELFARE...

...AND AN EXTREMELY **COHESIVE, COOPERATIVE SOCIAL SYSTEM**, CAPABLE OF QUICKLY DISSEMINATING NEW CULTURAL DEVELOPMENTS. THIS WOULD HAVE BEEN AN IMPORTANT ADVANTAGE WHEN IT CAME TO RAISING FAMILIES, HUNTING, AND GATHERING FOOD...

...AND **SURVIVING CONFLICTS** WITH OTHER SPECIES!

TO UNDERSTAND THE IMPACT OF THESE ADVANCES IN HUMAN DEVELOPMENT, WE NEED TO JUMP **FORWARD** TO A POINT IN TIME...

...WHERE THE PROTO-HUMAN TREE FORKED. THE PROTO-HUMAN LINEAGE DID NOT LEAD **SOLELY** TO MODERN HUMANS--

--**HOMO SAPIENS**-- BUT BRANCHED INTO OTHER SPECIES, INCLUDING *HOMO ERECTUS* AND *HOMO NEANDERTHALENSIS*.

BOTH OF THESE HUMAN SPECIES HAD MANAGED TO MIGRATE OUT OF AFRICA AS EARLY AS **1.8 MILLION YEARS AGO**, POPULATING EUROPE AND ASIA AS THEY FOLLOWED THEIR **OWN** EVOLUTIONARY PATHS.

ME GO NORTH.

ME WALK EAST.

IT WAS POSSIBLY THE HEAVILY MUSCLED, INTELLIGENT **NEANDERTHALS** WHO FRUSTRATED EARLY MODERN HUMANS' FIRST ATTEMPTS TO MIGRATE OUT OF AFRICA.

ARCHAEOLOGICAL EVIDENCE SUGGESTS THAT THEY WERE VERY EFFECTIVE HUNTERS OF BIG GAME.

SO HOW WAS IT THAT MODERN HUMANS EVENTUALLY **SUCCEEDED** IN OVERCOMING THE NEANDERTHALS, SUPPLANTING BOTH THEM AND *HOMO ERECTUS* ACROSS THE EARTH?

YEAH-- HOW THAT HAPPEN?

ONE POSSIBLE ANSWER LIES IN **FOXP2**, A RECENTLY DISCOVERED GENE, THE HUMAN VERSION OF WHICH HELPS CONTROL SOPHISTICATED MUSCLE MOVEMENTS IN THE LOWER FACE AND LARYNX, AS WELL AS AIDING IN SPEECH RECOGNITION.

IT WAS PERHAPS THE **FINAL REFINEMENT** THAT ALLOWED THE DEVELOPMENT OF THE COMPLEXITIES OF **HUMAN LANGUAGE**.

A... E... I... O... U...

?

FOXP2 WAS DISCOVERED THROUGH THE EXTREME **LINGUISTIC DISABILITIES** EXPERIENCED BY MEMBERS OF A CONTEMPORARY ENGLISH FAMILY.

GENETICISTS FOUND THAT ALL AFFECTED FAMILY MEMBERS HAD INHERITED A **DISABLED** FOXP2 GENE FROM A COMMON GRANDMOTHER.

FOX GENES ARE **ANCIENT**— MANY MAMMAL GENOMES, FROM MICE TO GORILLAS AND CHIMPANZEES, POSSESS VERSIONS.

BUT FOX GENES REMAINED **STABLE, CONSERVATIVE** GENES UNTIL THE **HUMAN VERSION** BEGAN AN **ACCELERATED EVOLUTION**...

HOOT.

WOOF.

MEOW.

SQUEAK.

...APPARENTLY NOT LONG AFTER THE CHIMP-HUMAN FORK IN THE PRIMATE FAMILY TREE OCCURRED. FOX GENES IN THE HUMAN GENOME **CONTINUED** THEIR MUTATIONS, WITH THE CURRENT VERSION DEVELOPING AS RECENTLY AS 200,000 YEARS AGO.

SOMEDAY OUR UNPARALLELED ORATORY SKILLS WILL WIN **HEARTS** AND **MINDS** AND SELL MANY USED CARS!

THE RAPIDLY EVOLVING VARIATION OF THE EARLY HUMAN FOX GENES WAS PERHAPS THE **KEY** FACTOR THAT ALLOWED THEM TO MASTER COMMUNICATION SKILLS FAR EXCEEDING THOSE OF ALL OTHER SPECIES.

NOW, TO RETURN TO THE EARLY HUMAN POPULATION'S **NEANDERTHAL** PROBLEM...

SOMETIME AROUND **50,000 YEARS AGO**, A SMALL BAND OF MODERN HUMANS FINALLY MANAGED TO MOVE OUT OF AFRICA AND SURVIVE WITHIN THE MIDST OF AN ESTABLISHED NEANDERTHAL POPULATION.

IN AN UNDOUBTEDLY RELATED EVENT, *HOMO NEANDERTHAL-ENSIS* AND *HOMO ERECTUS*, MODERN HUMANS' CLOSEST RELATIVES, DISAPPEARED COMPLETELY WITHIN THE NEXT **20,000 YEARS**!

THAT'S A RELATIVELY SHORT PERIOD OF TIME FOR TWO PREVIOUSLY ROBUST SPECIES TO VANISH. MODERN HUMANS' MIGRATIONAL SUCCESSES THROUGH THIS SAME TIME FRAME WOULD SEEM TO INDICATE SOMETHING MORE THAN COINCIDENCE AT WORK.

WHETHER THESE SPECIES' DISAPPEARANCES WERE **SINISTER** OR NOT MAY BE DISCOVERED THROUGH FURTHER GENETIC INVESTIGATION.

—THERE ARE **TWO** PREDOMINANT THEORIES CONCERNING THE **NEANDERTHAL**.

THEORY ONE

NEANDERTHALS WERE **OUT-COMPETED** BY MODERN HUMANS IN A BLOODY EXPANSIONIST **EXTERMINATION**. AS MODERN MAN GRADUALLY DISPLACED NEANDERTHALS ACROSS THE NORTHERN HEMISPHERE, THEY **DESTROYED** THEIR RIVALS, POPULATION BY POPULATION.

NEANDERTHAL REMAINS INDICATE THAT THEY WERE, **INDIVIDUALLY**, THE PHYSICALLY SUPERIOR SPECIES...

...BUT IF THEY DID **NOT** POSSESS THE SOPHISTICATED SOCIAL STRUCTURE AND LINGUISTIC SKILLS RECENTLY DEVELOPED BY **MODERN HUMANS**, THEY MAY HAVE REACHED A POINT OF SIGNIFICANT DISADVANTAGE.

?! NEW HUMAN SPEAR POINTS KILL BETTER AND BETTER...

AND HUMAN TEACH OTHER HUMAN TO MAKE NEW SPEAR SO **QUICKLY**!

THEORY TWO

THE NEANDERTHAL GENOME HAS BEEN **ABSORBED** INTO THE MODERN HUMAN GENOME THROUGH **INTERBREEDING**.

MAYBE NEANDERTHALS **DID** HAVE LINGUISTIC AND SOCIAL SKILLS THAT MODERN HUMANS FOUND ATTRACTIVE...

...AND *VICE VERSA*.

131

GENETIC ANTHROPOLOGISTS HOPE TO SOLVE THE MYSTERY WITH DNA THAT HAS BEEN RECENTLY EXTRACTED FROM A 38,000-YEAR-OLD NEANDERTHAL SKELETON.

OF PARTICULAR INTEREST AGAIN IS THE FOXP2 GENE.

IF THE **NEANDERTHAL** VERSION RESEMBLES THAT OF MODERN HUMANS, CHANCES ARE GREATLY INCREASED THAT NEANDERTHALS HAD COMPARABLE LINGUISTIC ABILITIES...

...CANCELING LANGUAGE AS A DISTINCT ADVANTAGE FOR THE MODERN HUMANS.

LOOK, WE COULD HURL DISPARAGING COMMENTS AT EACH OTHER ALL DAY OR WE COULD TRY AND GET BEYOND OUR DIFFERENCES.

OR, THE ULTIMATE **MAPPING** OF THE NEANDERTHAL GENOME COULD REVEAL MUCH **INTERMIXING** BETWEEN SPECIES— THE MODERN HUMAN GENOME MAY HAVE GOTTEN AS MUCH AS **25%** OF ITS DNA FROM NEANDERTHALS...

...INDICATING THAT NEANDERTHALS DID NOT VANISH AS THE RESULT OF VIOLENT COMPETITION...

...BUT WERE, THROUGH THE HAPPIER CIRCUMSTANCES OF INTERBREEDING, INSTEAD **LOST** AS A **SEPARATE** SPECIES AS THEIR GENOME WAS **INCORPORATED** INTO THE HUMANS'!

MAYBE TO BE HUMAN IS TO BE NEANDERTHAL!

WHATEVER THE CIRCUMSTANCES OF THEIR RIVALS' DISAPPEARANCE, MODERN HUMANS **DID** SUCCEED IN THE INCREDIBLY DIFFICULT TASK OF BREAKING OUT OF THE ANCESTRAL AFRICAN HOMELAND...

...AND SUCCESSFULLY SPREAD ACROSS THE REST OF THE PLANET!

A GOOD GUESS WOULD BE THAT HUMAN POPULATION DRIFT ACROSS THE CONTINENTS WAS PROBABLY PROPELLED BY PRESSURE FROM **WITHIN**.

WHEN A LOCATION BECAME STRESSED BY **OVERPOPULATION**, A **SEGMENT** OF THAT POPULATION WAS FORCED TO MOVE FORWARD TO NEW TERRITORIES.

GO FIND YOUR OWN PLACE!

"YOUR NEW ENSEMBLE LOOKS VERY **COMFORTABLE**..."

"IT FITS WELL, BUT FEELS LIKE SOMETHING'S **CRAWLING** THROUGH IT!"

TAILORING IS A LOUSY BUSINESS

MUCH OF EARLY HUMAN **CULTURAL INNOVATION** HAS LEFT NO ARCHAEOLOGICAL CLUES -- BUT GENETICS OFFERS ALTERNATE PATHS FOR REVEALING THE TIME LINE OF PREHISTORIC MAN'S ACHIEVEMENTS.

ANTHROPOLOGISTS HAVE LONG WONDERED EXACTLY WHEN IT WAS THAT HUMANS FIRST BEGAN TO TAILOR **CLOTHES** FOR THEMSELVES, DISPENSING WITH LOOSE, HANGING ANIMAL SKINS FOR MORE **FITTED** PROTECTION.

MARK STONEKING, A RESEARCHER IN THE FIELD OF EVOLUTIONARY ANTHROPOLOGY, DID SOME ADMIRABLE DETECTIVE WORK, SOLVING THAT MYSTERY BY LOOKING TO THE DNA OF A **PARASITICAL ORGANISM** AND APPLYING WHAT HE DISCOVERED TO DATING THE IMPORTANT INVENTION OF TAILORED CLOTHING.

MARK STONEKING

STONEKING'S INVESTIGATION BEGAN WITH HIS REALIZATION THAT THE COMMON HUMAN PARASITE, THE **HEAD LOUSE**, CAN SURVIVE ONLY WITHIN THE PROTECTION OF DENSE **HAIR** -- SO THE SPECIES IS RESTRICTED TO THE HAIRY HUMAN HEAD.

A **SECOND** LOUSE SPECIES, HOWEVER, **EVOLVED** FROM THE FIRST -- THE **BODY LOUSE**, WHICH EXPANDED ITS TERRITORY BY ADAPTING TO A LIFE SPENT **CLINGING** TO CLOTHING THAT FIT CLOSE TO THE BODY.

HEAD LOUSE

BODY LOUSE

IF THE TIME WHEN BODY LICE EVOLVED FROM HEAD LICE COULD BE FIXED, STONEKING REASONED, THEN IN ALL LIKELIHOOD THAT DATE WOULD ALSO INDICATE THE POINT WHEN HUMANS BEGAN TAILORING CLOTHES.

HE COLLECTED BOTH HEAD AND BODY LOUSE SPECIMENS FROM TWELVE SEPARATE POPULATIONS AROUND THE GLOBE. KNOWING THE RATE AT WHICH MUTATIONS WOULD ACCUMULATE WITHIN THE VARIOUS LICE GENOMES OVER CENTURIES, HE WAS ABLE TO COMPARE HIS SPECIMENS AND ULTIMATELY CALCULATE THE APPROXIMATE DATES OF EVOLUTIONARY FORKS IN THE LOUSE FAMILY TREE.

STONEKING DETERMINED THAT BODY LICE BRANCHED OUT FROM HEAD LICE AROUND **72,000 YEARS AGO**. ALTHOUGH LATER RESEARCH SUGGESTS THAT THAT BRANCHING MAY HAVE OCCURRED AS EARLY AS 107,000 YEARS AGO, IT IS A LOGICAL DEDUCTION THAT THE APPEARANCE OF THIS NEW SPECIES MUST HAVE QUICKLY FOLLOWED THE CREATION OF A NEW, EXPLOITABLE ECOLOGICAL NICHE -- **SEWN, FITTED CLOTHES**.

MODERN HUMANS APPARENTLY BECAME **FASHIONISTAS** WELL BEFORE THEY LEFT AFRICA. EVEN AT THAT EARLY DATE, THEIR **CULTURE** WAS AFFECTING THE EVOLUTIONARY COURSE OF OTHER SPECIES!

MOST INDIVIDUALS POPULATING A NEW TERRITORY, HOWEVER, TENDED TO **STAY** IN THAT PLACE WHERE THEY HAD SET DOWN ROOTS, AND SO DID THE VAST MAJORITY OF THEIR DESCENDANTS.

GENERATIONS OF STABLE, ISOLATED POPULATIONS EVOLVED THEIR OWN **DISTINCT VARIATIONS** ON THE HUMAN GENOME.

THE TALL TREES BEAR DELICIOUS FRUIT, BUT ONLY **I** CAN REACH THEM!

EACH POPULATION'S GENOME WAS DEFINED THROUGH REGIONAL **ENVIRONMENTAL PRESSURES, RANDOM MUTATION,** AND **NATURAL SELECTION.**

WHEN THE EXCHANGE OF MUTATIONS-- **ALLELES** -- IS LIMITED TO A SPECIFIC POPULATION, THAT POPULATION DEVELOPS **"SIGNATURE"** GENETIC TRAITS, IDIOSYNCRATIC TRAITS PARTICULAR TO THAT POPULATION.

GEOGRAPHIC ISOLATION AND THE TENDENCY TO "STAY HOME" **REINFORCE** THOSE TRAITS WITHIN THE POPULATION.

WHY, YES, I **DO** COME FROM THE LAND OF THE TALL TREES...

ENVIRONMENTAL FACTORS ARE PERHAPS MOST OFTEN THE FORCES BEHIND GENOME VARIATION WITHIN POPULATIONS.

FOR INSTANCE...

...DARK SKIN AND HAIR OFFER THE GREATEST PROTECTION AGAINST SOLAR RADIATIONS AND ARE FOUND MOST COMMONLY IN POPULATIONS LOCATED CLOSEST TO EARTH'S **EQUATOR,** WHERE THE SUN'S RAYS ARE THE MOST **DIRECT...**

...WHILE IN REGIONS CLOSER TO THE **POLES,** WHERE SUNLIGHT IS AT A **PREMIUM,** GENOMES FEATURING PALE SKIN AND HAIR, WHICH ABSORB SUNLIGHT MORE READILY, ARE MORE COMMON.

A MORE SPECIFIC AND COMPLEX EXAMPLE OF **POPULATION GENETICS** WOULD BE THAT OF **LACTOSE TOLERANCE** --

-- THE ABILITY TO DIGEST **MILK.**

AS HUMAN SOCIETIES **DIVERSI-FIED** -- PARTIALLY AS A RESULT OF THE ABILITY TO COMMUNI-CATE WITH SOPHISTICATION-- SOME POPULATIONS GAVE UP THE HUNTER-GATHERER NO-MADIC EXISTENCE IN FAVOR OF SETTLED AGRICULTURE.

PEOPLES IN EUROPE AND AFRICA WHO DOMESTICATED **CATTLE** FOUND IT AN ADVAN-TAGE TO INCORPORATE THEIR STOCK'S MILK INTO THEIR DIET.

MILK IS FULL OF **CALORIES** -- MOST OF WHICH COME FROM THE SUGAR **LACTOSE**. THE ENZYME NECESSARY FOR DIGESTING LACTOSE IS **LACTASE**, WHICH IS CODED BY A GENE THAT IS GENERALLY **SWITCHED OFF** WHEN AN INDIVIDUAL IS WEANED.

MILK IS IMPORTANT FOR INFANT GROWTH BUT IS NOT A RESOURCE NATURALLY AVAILABLE TO ADULTS, SO THE PRODUCTION OF LACTASE BECOMES **EXPENDABLE**.

CATTLE-HERDING PEOPLES IN AFRICA AND EUROPE DEVELOPED **MUTATIONS** THAT KEPT THE LACTASE PRODUCTION GENE **SWITCHED PERMANENTLY ON**. THEY PROSPERED WITH THEIR MILK-SUPPLEMENTED DIET AND SO HAD A BETTER CHANCE OF REPRODUCING AND PASSING ON THEIR NEW ALLELE TO THEIR OFFSPRING.

OF COURSE, IF THE LACTOSE-TOLERANCE ALLELE APPEARED IN POPULATIONS THAT WERE **NOT** HERDERS, IT WAS OF **NO** ADVANTAGE.

ANOTHER EXAMPLE OF POPULATION GENETICS ENTAILS PROTECTIONS AGAINST **MALARIA**, A TROPICAL DISEASE THOUGHT TO HAVE BECOME A DEADLY THREAT IN RELATIVELY RECENT HUMAN HISTORY...

...AS HUMAN AGRICULTURAL CULTIVATION OPENED UP VAST BREEDING HABITATS FOR **MOSQUITOES** CARRYING THE MALARIA PARASITE.

MUTATIONS THAT ALTER HEMOGLOBIN AND PROVIDE A **PROTECTION** AGAINST MALARIA WERE SELECTED IN INDEPENDENT POPULATIONS IN **AFRICA** AND IN POPULATIONS BORDERING THE **MEDITERRANEAN SEA**.

THEY SAY THE BLOOD TASTED **DIFFERENT** BACK IN GREAT-GRANDDAD'S DAY.

IN A CASE OF UNFORTUNATE **PLEIOTROPY**, HOWEVER...

...THOSE NEW ALLELES RESPONSIBLE FOR HINDERING MALARIA **ALSO** GAVE RISE TO VARIOUS BLOOD DISEASES...

...MOST NOTORIOUSLY, SICKLE-CELL ANEMIA.

THE APPEARANCE OF BOTH **LACTOSE TOLERANCE** AND **MALARIA RESISTANCE** ILLUSTRATES SOME IMPORTANT ASPECTS OF HUMAN EVOLUTION:

1) THE HUMAN GENOME IS **NOT A FIXED** THING -- IT IS CONSTANTLY **CHANGING.**

2) WIDELY SEPARATED POPULATIONS, RESPONDING TO **SIMILAR CONDITIONS,** CAN **INDEPENDENTLY** ARRIVE AT **SIMILAR GENETIC SOLUTIONS.**

YOU MEAN TO SAY THAT **YOU** GOT MILK, TOO?

3) GENES CAN RESPOND TO **ENVIRONMENTAL** CONDITIONS, INCLUDING **CULTURAL** AND **BEHAVIORAL** FACTORS.

SEEMS THAT THERE ARE MORE AND MORE PESKY MOSQUITOES FOR EVERY HECTARE WE CLEAR!

YEAH, BUT AT LEAST WE'RE NOT GETTING AS SICK AS OUR FATHERS' GENERATION DID...

OF COURSE, THE GENOMES OF ALL OTHER ORGANISMS ARE **LIKEWISE** CONSTANTLY MUTATING, ADAPTING TO CHANGING ENVIRONMENTAL CONDITIONS -- **INCLUDING** THOSE FOR WHICH HUMANS ARE RESPONSIBLE.

EVERY SPECIES **SUCCESSFULLY EVOLVES** -- OR **FAILS** -- IN RESPONSE TO ALL THE OTHER SPECIES WITH WHICH IT COMES INTO CONTACT!

SWAT!!

I DON'T KNOW **HOW** WE MANAGED TO SURVIVE BEFORE THESE MAMMALIAN BIPEDS CAME ALONG!

YIKES! What you've just covered gives me **HOPE** --

-- maybe somewhere here on Glargal there are **SQUINCH** with mutations that are even now countering our genetic disorder...

THAT'S A **POSSIBILITY,** ALTHOUGH WITHOUT **SEXUAL** REPRODUCTION, THE DISPERSION OF ANY HELPFUL MUTATION WOULD BE AWFULLY SLOW.

BUT YOUR THOUGHTS ARE HEADED IN THE RIGHT DIRECTION, MOST SAGACIOUS SIRE!

ON EARTH, THE HUMANS ARE NOW LOOKING AT THE GENETIC IDIOSYNCRASIES OF SPECIFIC POPULATIONS TO DETERMINE WHAT CAN BE LEARNED OF EACH POPULATION'S SUCCESSES AND LIABILITIES REGARDING DISEASES AND DISORDERS.

THE COMPLETE SEQUENCING OF HUMAN DNA BY THE HUMAN GENOME PROJECT HAS MADE POSSIBLE THE **INTERNATIONAL HAPMAP PROJECT**, WITH THE AMBITIOUS GOAL OF DEVELOPING A **HAPLOTYPE MAP** OF THE HUMAN GENOME.

International HapMap Project

A **HAPLOTYPE** IS AN EXTENDED SEQUENCE OF DNA, COMPRISING MANY GENES, THAT HAPPENS TO BE INHERITED **AS A BLOCK** WITHIN A GIVEN POPULATION.

HAPLOTYPE BLOCK

GENE   GENE   GENE

OF COURSE, WITHIN ANY HAPLOTYPE, THERE WILL BE MUCH **ALLELIC VARIATION.**

HAPMAP RESEARCHERS COMPARE HAPLOTYPE BLOCKS WITHIN A POPULATION TO TRY TO DETERMINE WHICH ALLELIC VARIATIONS MAY LEAD TO A GREATER **RESISTANCE**, OR A GREATER **SUSCEPTIBILITY**, TO CERTAIN DISEASES...

...AS WELL AS WHICH MAY ASSIST A GREATER OR LESSER RESPONSIVENESS TO SPECIFIC **DRUGS** OR **VACCINES**.

A BROAD GOAL OF THE HAPMAP PROJECT IS TO FIND **PATTERNS** OF HUMAN GENETIC VARIATION -- TO LINK GENES AFFECTING HEALTH TO SPECIFIC POPULATIONS AND ENVIRONMENTAL AND CULTURAL FACTORS...

...WITH THE ULTIMATE HOPE BEING THAT A **CATALOG** OF GENETIC VARIATION WILL ALLOW THE MEDICAL PROFESSION TO CUSTOMIZE MORE EFFECTIVE TREATMENTS -- IDEALLY, RIGHT DOWN TO TREATING A SPECIFIC **INDIVIDUAL**.

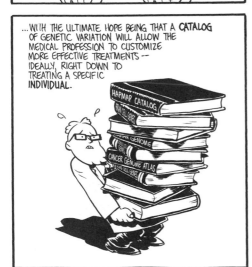

AND **THAT** IS WHY A THOROUGH UNDERSTANDING OF ANY **INDIVIDUAL'S** GENETIC HISTORY, AS WELL AS ANY **SPECIES'**, IS IMPORTANT:

WHILE THE HUMAN GENOME PROJECT CREATED A WORKING **COMPOSITE** GENOME OF THE MODERN HUMAN, THE VARIATIONS **WITHIN** THAT GENOME -- THE MUTATIONS THAT PROVIDE SUCH DIVERSITY-- MUST **ALL** BE RECOGNIZED...

... IF THE MOST EFFECTIVE REMEDIES POSSIBLE -- THOSE DESIGNED EXPRESSLY FOR THE **INDIVIDUAL**, WHETHER TO TREAT MALADIES THAT ARE **DIRECTLY** THE RESULT OF GENETIC DISORDERS...

DOWN SYNDROME

NORMAL RED BLOOD CELLS

SICKLE-CELL-ANEMIA-AFFLICTED RED BLOOD CELLS

... OR FOR THOSE MADE MORE **LIKELY** BY GENETIC VARIATIONS -- ARE TO BECOME A REALITY.

VERY RECENTLY, **JAMES WATSON** -- ONE MEMBER OF THE TEAM WHO DISCOVERED THE MOLECULAR STRUCTURE OF DNA -- HAD **HIS** PERSONAL GENOME SEQUENCED...

...AT A COST OF **UNDER $1 MILLION.**

THAT'S A SIGNIFICANT MONETARY FIGURE, IF SOMEWHAT SYMBOLIC. PREVIOUSLY, IT HAD COST **MANY MILLIONS** OF DOLLARS TO SEQUENCE A GENOME.

GENOME SEQUENCING *now only* $999,999⁹⁹!

BUT CONTINUING BREAKTHROUGHS IN SEQUENCING TECHNOLOGIES PROMISE THE **FURTHER** REDUCTION OF THAT EXPENSE.

THERE IS HOPE THAT THE COST OF PERSONAL GENOME SEQUENCING CAN BE REDUCED TO A FEW **THOUSAND** DOLLARS AND SO BECOME AVAILABLE TO **ALL** HUMAN PATIENTS...

... PROVIDING VERY SPECIFIC MAPS FOR THE CARE OF THE **INDIVIDUAL**...

... AS OPPOSED TO THE VERY **GENERALIZED** CARE NOW AVAILABLE.

WHAT IS PAST   IS NOT FORGOTTEN

?

POPULAR ENTERTAINMENT HAS CREATED THE WIDESPREAD NOTION THAT ANCIENT, EXTINCT CREATURES MIGHT BE **RAISED FROM THEIR GRAVES** VIA CLONING IF ONLY ADEQUATE SAMPLES OF THEIR DNA COULD BE FOUND. THE REALITY IS THAT DNA DECAYS PRETTY RAPIDLY AFTER AN ORGANISM'S DEATH. IN FACT, UNTIL RECENTLY, IT WAS ACCEPTED THAT **ALL** THE SOFT TISSUE OF ANY ORGANISM THAT ESCAPED COMPLETE DECOMPOSITION AFTER DEATH WOULD BE **FOSSILIZED** AND REPLACED BY **MINERALS** WITHIN A **MILLION YEARS**.

SO IT CAME AS QUITE A SHOCK TO THE SCIENTIFIC COMMUNITY WHEN PALEONTOLOGIST **MARY SCHWEITZER** ANNOUNCED THAT SHE'D EXTRACTED SOFT TISSUE FROM THE REMAINS OF A FEMUR BONE BELONGING TO A *TYRANNOSAURUS REX* FOSSIL OVER **68 MILLION YEARS OLD**!

MARY SCHWEITZER

THE SOFT TISSUE FROM THE DINOSAUR WAS A **COLLAGEN** -- A FIBROUS MATERIAL THAT GIVES BONES STRUCTURE AND FLEXIBILITY, AND IS COMPOSED OF **PROTEINS**. PROTEIN ANALYSIS REVEALED THAT THE T. REX SEQUENCES MATCHED BETTER TO THE CONTEMPORARY **CHICKEN** THAN TO ANY OTHER SINGLE ORGANISM OR CURRENTLY KNOWN SEQUENCE. THIS DISCOVERY ADDS SIGNIFICANT WEIGHT TO THE THEORY THAT BIRDS ARE INDEED THE LIVING DESCENDANTS OF THE DINOSAURS.

WITH THE **CHICKEN GENOME** HAVING BEEN COMPLETELY SEQUENCED (THANKS TO THE HUMAN GENOME PROJECT), IT CAN JUSTIFIABLY BE WONDERED WHAT DINOSAURIAN GENES MAY LIE DISCARDED BUT STILL INTACT WITHIN THAT GENOME'S "JUNK" DNA, WAITING TO BE IDENTIFIED. MUCH OF THE 250-MILLION-YEAR-OLD HISTORY OF DINOSAURS MIGHT BE REVEALED WITHIN A BARNYARD FOWL!

DINOSAURS WON'T BE CLONED FROM THEIR FOSSIL REMAINS, BUT THE RESURRECTION OF A MORE **RECENTLY** EXTINCT CREATURE **MAY** BE POSSIBLE. THE **WOOLLY MAMMOTH** ROAMED THE NORTHERN HEMISPHERE AS RECENTLY AS **11,000 YEARS AGO**. EXPERTS IN ARTIFICIAL FERTILIZATION **KAZUFUMI GOTO** AND **AKIRA IRITANI** OF THE **MAMMOTH CREATION PROJECT** HOPE TO FIND SUFFICIENTLY PRESERVED DNA WITHIN ONE OF THE FROZEN MAMMOTH CARCASSES OCCASIONALLY UNCOVERED FROM THE SIBERIAN TUNDRA.

IF THEY CAN EXTRACT DNA THAT HAS NOT BEEN TOO **FRAGMENTED** BY DECAY-- **A BIG** IF -- THEY PLAN TO FERTILIZE THE EGG OF AN **ASIAN ELEPHANT** WITH MAMMOTH SPERM. THEIR HOPE IS THAT, FOLLOWING A MULTI-GENERATIONAL BREEDING PROGRAM, A CREATURE **88%** WOOLLY MAMMOTH WILL BE PRODUCED WITHIN 50 YEARS.

A SPECTACULAR ACHIEVEMENT, TO BE SURE, BUT THE ARTIFICIAL MAMMOTH WOULD BE AN ORGANISM CREATED **OUT OF CONTEXT** -- SEPARATED FROM THE ENVIRONMENT IN WHICH THE SPECIES HAD EVOLVED AS A WORKING COMPONENT-- AND PROBABLY SUBJECT TO DISEASES AGAINST WHICH IT HAS NO DEFENSES.

139

AND THAT...

...UMM...

... **THAT** CONCLUDES MY REPORT ON GENETICS FROM THE PLANET EARTH...

Hmm.

As you chose to end by emphasizing the importance that humans place on understanding genetics on an individual basis...

...I assume you are suggesting that this is the path **WE** follow?

ONLY --GULP-- ONLY...

...**ONLY** BECAUSE IT IS IN THE ROYAL FAMILY'S **BEST INTERESTS** TO HAVE AS MANY GENETIC OPTIONS AVAILABLE AS POSSIBLE.

IF THE REST OF SQUINCH-DOM **ALSO** HAPPENS TO BENEFIT, SO MUCH THE BETTER.

IF OUR GENETIC ORGANIZATION IS **ANYTHING** LIKE THAT ON EARTH, THE MORE WE LEARN ABOUT **EACH OTHER**...

... THE MORE WE WILL HELP **OURSELVES!**

AND THE MORE WE UNDERSTAND OF OUR EVOLUTIONARY **HISTORY**, THE BETTER GRIP WE'LL HAVE ON OUR **CURRENT** CRISIS...

OKAY, I **GET** IT.

WE MUST--

OKAY, *OKAY!* You've *SOLD* me, Bloort!

LISTEN UP, EVERYBODY!

I'm making a *GLOBAL DECLARATION OF INTENT,* by the power vested in me as *LORD OF ALL SQUINCHDOM!*

On this day, we launch a *TOTAL SCIENTIFIC COMMITMENT* to uncover our genetic heritage!

We'll learn all about our *MOLECULAR, CELLULAR,* and *TRANSMISSION* strategies!

We will even investigate the possibility of improving our genetic options by developing technologies that will allow us to -- *ugh* -- *SWAP GENES.*

AND, Bloort 183 -- we will research our *GENETIC HISTORY...*

...*EVERY* Squinch's genetic history...

...applying that knowledge to develop therapies that'll *BEAT* this disorder!

MIGHT I SUGGEST, YOUR MOST GRACIOUS OMNIPOTENCE, THAT MY UNIQUE EXPERIENCE WOULD MAKE ME THE *PERFECT* CHOICE TO HEAD UP OUR--

NONSENSE, Bloort! You're *FAR* too valuable on that fascinating planet Earth, collecting more information! I want you back there *IMMEDIATELY!*

LET'S GET TO WORK, FOLKS!

Where's my speechwriter?!

IT IS A
FASCINATING
PLANET.

Suggested Reading
Glossary

SUGGESTED READING

PERIODICALS

SCIENTIFIC AMERICAN (www.SciAm.com)

NATURAL HISTORY (www.naturalhistorymag.com)

NATIONAL GEOGRAPHIC (www.nationalgeographic.com)

DISCOVER (www.discovermagazine.com)

NEW SCIENTIST (www.newscientist.com)

SCIENCE NEWS (www.sciencenews.org)

BOOKS

Dawkins, Richard. *THE SELFISH GENE*, 3rd ed. New York: Oxford University Press, 2006.

Fortey, Richard. *LIFE: A NATURAL HISTORY OF THE FIRST FOUR BILLION YEARS OF LIFE ON EARTH.* New York: Vintage Books, 1999.

Fritz, Sandy, ed. *UNDERSTANDING CLONING: FROM THE EDITORS OF SCIENTIFIC AMERICAN.* New York: Warner Books, 2002.

Hartl, Daniel L., and Elizabeth W. Jones. *GENETICS: ANALYSIS OF GENES AND GENOMES*, 6th ed. Boston: Jones and Bartlett, 2004.

Hartwell, Leland, et al. *GENETICS: FROM GENES TO GENOMES*, 3rd ed. New York: McGraw-Hill, 2006.

Pollan, Michael. *THE BOTANY OF DESIRE.* New York: Random House, 2001.

Wade, Nicholas. *BEFORE THE DAWN: RECOVERING THE LOST HISTORY OF OUR ANCESTORS.* New York: Penguin, 2006.

Watson, James D., and Berry Andrew. *DNA: THE SECRET OF LIFE.* New York: Knopf, 2004.

WEBSITES

GENOME NEWS NETWORK
www.genomenewsnetwork.org

HOWARD HUGHES MEDICAL INSTITUTE
www.hhmi.org

HUMAN GENOME PROJECT INFORMATION
www.ornl.gov/sci/techresources/Human_Genome/home.shtml

NATIONAL CENTER FOR BIOTECHNOLOGY INFORMATION
www.ncbi.nlm.nih.gov

NATIONAL HUMAN GENOME RESEARCH INSTITUTE
www.genome.gov

PALOMAR COLLEGE (for specific information about Mendelian genetics)
anthro.palomar.edu/mendel

UNIVERSITY OF CALIFORNIA BERKELEY (for specific information about evolution)
evolution.berkeley.edu

GLOSSARY

ADENINE: Nucleic acid of both DNA and RNA.

ADENOVIRAL VECTORS: Transmission of an infectious agent to a host cell using the viral coat and molecular mechanisms of an Adenovirus.

ADENOVIRUSES: A group of DNA-containing viruses that cause upper respiratory tract infections in humans.

ALLELES: Gene variants with the same locus on homologous chromosomes.

AMINO ACIDS: Building blocks of proteins; they contain amino and carboxylic acid groups.

ANAPHASE: Phase of cell division during which sister chromatids separate into independent chromosomes and the cleavage furrow forms.

ANAPHASE

ANEUPLOIDY: An abnormal chromosome number that is not an exact multiple of the haploid number.

ANTI-CODON: A sequence of three adjacent nucleotides that bonds to a corresponding codon in messenger RNA (mRNA) during protein synthesis.

AUTOSOMAL DOMINANT: A phenotype resulting from the dominant expression of one allele over the other in a gene pair.

AUTOSOMAL RECESSIVE: A phenotype in which a homozygous recessive genotype is expressed.

AUTOSOMES: All chromosomes that do not serve to determine the sex of an individual organism.

AUTOTROPHS: Organisms that are able to independently synthesize organic molecules from inorganic material and/or convert solar energy into usable energy.

AUTOTROPH

BACTERIA: A single-celled organism that lacks a defined nuclear membrane.

BRCA 1 GENE: Breast Cancer 1 Gene—a tumor suppressor gene, commonly mutated in females with breast and/or ovarian cancer, that normally serves to repair damaged DNA.

CANCER: A group of diseases similarly characterized by uncontrolled cellular replication.

CENTRIOLES: Organelles involved in spindle formation during the cell-division stage called anaphase.

CENTROMERE: The region at which chromatids are joined and to which spindle fibers attach during cell division.

CHROMATID: Either of two daughter strands of a duplicated chromosome that are joined by a centromere and separate during cell division to become individual chromosomes.

CHROMATID

CHROMOSOME: A linear arrangement of tightly coiled DNA located in a cell's nucleus.

CLONING: Duplication of a biological entity.

CODOMINANCE: The result of nonidentical alleles at a heterozygous gene locus expressing fully as both phenotypes.

CODON: A specific sequence of three adjacent nucleic acids on a strand of DNA that code for an amino acid.

CRI DU CHAT SYNDROME: A congenital disease resulting from a partial deletion of chromosome 5, characterized by mental retardation and a catlike cry.

CYSTIC FIBROSIS: An autosomal recessive disease caused by a mutation in the cystic fibrosis transmembrane conductance regulator (CFTR) gene.

CYTOKINESIS: Separation of the cytoplasm of the parent cell as two daughter cells form during cell division.

CYTOPLASM (aka protoplasm): The fluid substance that fills the cell.

CYTOSINE: Nucleic acid of both DNA and RNA.

DIPLOID: Cells and/or organisms having two copies of each chromosome.

DNA: Deoxyribonucleic acid. It carries the genetic information in the cell and is capable of self-replication..

DNA

DOWN SYNDROME: A developmental disorder in humans characterized by mental retardation resulting from an extra copy of chromosome 21.

DUCHENNE MUSCULAR DYSTROPHY: An X-linked recessive disease caused by a defective gene for dystrophin and resulting in progressive muscle atrophy and weakness.

ENDOPLASMIC RETICULUM: A membrane network within the cytoplasm of cells involved in synthesis, modification, and transport of cellular material.

EPISTASIS: Suppression of one gene by another unrelated gene.

EUKARYOTE: An organism composed of cells containing a nucleus with linear chromosomes contained within a nuclear membrane, and which replicates by mitosis.

EUKARYOTE CELL

EXONS: Sequences of DNA that code for proteins.

FRAGILE X SYNDROME: An X-linked dominant disease caused by a mutation in the FMR1 gene resulting in mental retardation and autism.

FUNGUS: Any organism from the kingdom Fungi, including yeasts, molds, and mushrooms.

FUNGUS

GEMMULE: The bud of an asexually reproducing parent organism that becomes its offspring.

GENE: A heritable unit of DNA that encodes or directly contributes to protein synthesis.

GENETIC DRIFT: Reduction of heterozygosity within a population due to the random reduction of an allele in the gene pool over time; inversely related to the population size.

GENOME: The entire complement of genes in a species.

GENOTYPE: The entire genetic constitution of an individual.

GONOSOMES: Sex-determining chromosomes.

GUANINE: Nucleic acid of both DNA and RNA.

GYRASE: An enzyme that increases the density of the DNA helix by introducing negative supercoils and removing positive supercoils.

HAPLOID: A cell or organism having only one copy of each chromosome.

HAPLOTYPE: A group of closely related genes often inherited as a unit.

HAPLOID ORGANISM

HELICASE: An enzyme that aids in the separation of double-stranded DNA, thereby allowing single-stranded DNA to be copied.

HETEROTROPHS: Organisms that rely on other organisms for their energy requirements.

HETEROZYGOUS: Term applied to alleles in a gene pair that are different at a given chromosomal locus.

HISTONES: Proteins in a cell's nucleus around which DNA tightly coils.

HOLOENZYME COMPLEX: A complex of approximately 20 proteins, including transcription factors, that binds to the promoter region to initiate transcription.

HOMOLOGOUS: Similar in structure and evolutionary origin, though not necessarily in function.

HOMOZYGOUS: Term applied to identical alleles in the gene pair at a given chromosomal locus.

HUNTINGTON'S DISEASE: An autosomal dominant disease caused by a mutation in the Huntington (Htt) gene, leading to an alteration in the Htt protein and neuronal death.

INCOMPLETE DOMINANCE: Partial expression of non-identical alleles resulting in an intermediate phenotype.

PP   Pp   pP   PP

INCOMPLETE DOMINANCE

INCOMPLETE PENETRANCE: A condition, due to environmental conditions or the interactions of non-allelic genes, where the genotype is not fully realized in the phenotype.

INCONTINENTIA PIGMENTI: An X-linked dominant disease caused by mutations in the NEMO gene and resulting in multiple disorders of the skin and nervous system.

INDEPENDENT ASSORTMENT: A Mendelian law stating that maternal and paternal chromosomes are bequeathed to gametes independently of one another during meiosis.

INTRONS: Non-coding DNA that exists between exons.

KLINEFELTER'S SYNDROME: A chromosomal anomaly in males affecting sexual development and resulting in sterility.

LENTIVIRAL VECTORS: Means of transmission of an infectious agent to a host cell using the viral coat and molecular mechanisms of a lentivirus.

LENTIVIRUSES: A subclass of retrovirus that is able to infect non-dividing cells as well as dividing cells.

LIGASE: An enzyme that joins unattached ends of DNA.

LYSOSOME: A small cavity in the cell's cytoplasm containing enzymes involved in protein degradation and natural cell death.

MARFAN SYNDROME: An autosomal dominant disorder with multiple defects caused by mutations in the gene coding for fibrilin, a protein of the connective tissue.

MEIOSIS: The process by which gametes--male and female sex cells--are formed.

MESSENGER RNA (mRNA): RNA that mediates the transfer of genetic information from a cell nucleus to ribosomes in the cytoplasm, where it serves as a template for protein synthesis.

METAPHASE: The second stage of cell division in which the chromosomes are arranged in the equatorial plane prior to separation.

MITOCHONDRION

MITOCHONDRIA: Organelles in the cytoplasm of eukaryote cells important for cell metabolism.

MITOSIS: The process of cellular and genetic replication.

MUTATION: An inconsistency in an organism's genetic code.

NUCLEOSOME: A protein core of histones around which DNA tightly coils.

ONCORETROVIRAL VECTOR: Means of transmission of an infectious agent to a host cell using the viral coat and molecular mechanisms of an oncoretrovirus.

ONCORETROVIRUS: A retroviral virus commonly used as a vector for gene therapy.

PARKINSON'S DISEASE: A disease often caused by insufficient production and/or action of dopamine; characterized by progressive degeneration of the nervous system.

PARTHENOGENESIS: Development of an unfertilized gamete into an adult organism.

PHENOTYPE: Observed genetic expression of an organism's physical, biochemical, and physiological characteristics.

PLEIOTROPY: The expression of multiple phenotypes from a single gene.

POLYMERASE: An enzyme that aids in the attachment of nucleic acids to one another during the transcription process.

POLYPEPTIDE: A polymer of amino acids linked by peptide bonds; may be a protein fragment.

POLYSACCHARIDES: Carbohydrates composed of simple sugars.

PROGESTERONE: A hormone that acts to prepare the uterus for implantation of a fertilized ovum and helps maintain a healthy pregnancy.

PROKARYOTE: A single-celled organism that lacks a membrane-bound nucleus.

PROKARYOTE

PROPHASE: In cell division, the stage where centrosomes move to opposite poles and spindle fibers form.

PROTEINS: Polymers of amino acids linked by peptide bonds.

PROTIST: Single-celled, eukaryotic organisms, including protozoans, slime molds, and certain algae.

RETROVIRUS: Any virus with an RNA genome that, following entry into a host cell, is reverse transcribed into double-stranded DNA and integrates into the host genome.

RETROVIRUS

RIBOSOMAL RNA (rRNA): A class of RNA found in the ribosomes of cells involved in protein synthesis.

RIBOSOMES: Organelles involved in protein synthesis.

RIBOSOME

RNA: Ribonucleic acid. A single-stranded molecule that serves as the intermediary between DNA and protein production.

SEGREGATION: Separation of paired alleles on homologous chromosomes during meiosis, ensuring each gamete does not receive duplicate alleles.

STEM CELL: A cell that is not completely differentiated and retains the ability to produce daughter cells that can further differentiate.

STROMATOLITES: Fossilized layers of cyanobacteria, which are also known as blue-green algae.

TAY-SACHS DISEASE: An autosomal recessive disease ultimately resulting in neuronal death.

TELOMERASE: An enzyme responsible for the addition of telomeres to chromosome ends.

TELOMERE: A single-stranded, non-coding molecular cap on the ends of eukaryotic chromosomes that prevents joining with chromosome fragments.

TELOPHASE: The final stage of cell division, during which the chromosomes arrive at the poles of the cell and the cytoplasm divides.

THYMINE: Nucleic acid of DNA.

TOTIPOTENCE: The quality of a cell that allows it to retain the ability to differentiate into any cell type.

TRANSCRIPTION: The process by which mRNA is synthesized from a DNA template, resulting in the transfer of genetic information from the DNA molecule to the mRNA.

TRANSFER RNA (tRNA): RNA molecule responsible for transferring amino acids to the ribosomes for protein assembly.

TRANSFER RNA

TRANSLATION: The process by which nucleic acids of DNA are recoded into amino acids for protein synthesis.

URACIL: Nucleic acid of RNA.

VECTOR: An organism or virus that transmits an infectious agent from one host organism to another.

VIRUS: Transmissible DNA or RNA enveloped by glycoprotein and requiring a host organism for replication.

ZYGOTE: The term applied to the diploid cell resulting from the union of a male and female gamete until the first cleavage.

ZYGOTE

ABOUT THE SCRIPTER

MARK SCHULTZ has been cartooning, illustrating, and writing for longer than he cares to admit. His best-known creation is the award-winning speculative adventure comic book *Xenozoic Tales*, which ran for fourteen issues and has been adapted to television as the animated series *Cadillacs and Dinosaurs*. He has also created and written the undersea adventure *SubHuman* with paleontologist Michael Ryan. In addition to his own works, Schultz has drawn or scripted many popular fictional icons, including Superman, Flash Gordon, Tarzan, the Spirit, and Conan of Cimmeria. Currently, he writes the syndicated Sunday comic strip *Prince Valiant*, and continues producing material for his *Various Drawings* book collections. His most recently published work is the illustrated novella *Storms at Sea*. Schultz's scripts and illustrations have garnered five Harvey Awards, two Eisners, and an Inkpot. He lives in northeastern Pennsylvania.

ABOUT THE ILLUSTRATORS

Unrelated cartoonists ZANDER CANNON and KEVIN CANNON have been together as a studio since 2004, but their work in comics stretches back to 1993, with such titles as *The Replacement God* (Slave Labor Graphics and Image Comics, 1995), *Top Ten* (America's Best Comics, 1999), and *Smax* (America's Best Comics, 2003). They have also done various work for DC Comics, Dark Horse Comics, the National Oceanic and Atmospheric Administration, and the University of Nagoya. Their studio work includes the graphic novel *Bone Sharps, Cowboys, and Thunder Lizards* (G.T. Labs, 2005) and a number of other comics, as well as illustration, design, and animation work. They live and work in Minneapolis, Minnesota, and claim to be card-carrying members of the International Cartoonist Conspiracy.